Line Integral Methods for Conservative Problems

MONOGRAPHS AND RESEARCH NOTES IN MATHEMATICS

Series Editors

John A. Burns
Thomas J. Tucker
Miklos Bona
Michael Ruzhansky

Published Titles

Application of Fuzzy Logic to Social Choice Theory, John N. Mordeson, Davender S. Malik and Terry D. Clark

Blow-up Patterns for Higher-Order: Nonlinear Parabolic, Hyperbolic Dispersion and Schrödinger Equations, Victor A. Galaktionov, Enzo L. Mitidieri, and Stanislav Pohozaev

Cremona Groups and Icosahedron, Ivan Cheltsov and Constantin Shramov

Difference Equations: Theory, Applications and Advanced Topics, Third Edition, Ronald E. Mickens

Dictionary of Inequalities, Second Edition, Peter Bullen

Iterative Optimization in Inverse Problems, Charles L. Byrne

Line Integral Methods for Conservative Problems, Luigi Brugnano and Felice Iavernaro

Lineability: The Search for Linearity in Mathematics, Richard M. Aron, Luis Bernal González, Daniel M. Pellegrino, and Juan B. Seoane Sepúlveda

Modeling and Inverse Problems in the Presence of Uncertainty, H. T. Banks, Shuhua Hu, and W. Clayton Thompson

Monomial Algebras, Second Edition, Rafael H. Villarreal

Nonlinear Functional Analysis in Banach Spaces and Banach Algebras: Fixed Point Theory Under Weak Topology for Nonlinear Operators and Block Operator Matrices with Applications, Aref Jeribi and Bilel Krichen

Partial Differential Equations with Variable Exponents: Variational Methods and Qualitative Analysis, Vicenţiu D. Rădulescu and Dušan D. Repovš

A Practical Guide to Geometric Regulation for Distributed Parameter Systems, Eugenio Aulisa and David Gilliam

Signal Processing: A Mathematical Approach, Second Edition, Charles L. Byrne

Sinusoids: Theory and Technological Applications, Prem K. Kythe

Special Integrals of Gradshetyn and Ryzhik: the Proofs – Volume I, Victor H. Moll

Forthcoming Titles

Actions and Invariants of Algebraic Groups, Second Edition, Walter Ferrer Santos and Alvaro Rittatore

Analytical Methods for Kolmogorov Equations, Second Edition, Luca Lorenzi

Complex Analysis: Conformal Inequalities and the Bierbach Conjecture, Prem K. Kythe

Forthcoming Titles (continued)

MONOGRAPHS AND RESEARCH NOTES IN MATHEMATICS

Line Integral Methods for Conservative Problems

Luigi Brugnano
Università di Firenze, Italy

Felice Iavernaro
Università di Bari, Italy

CRC Press
Taylor & Francis Group
Boca Raton London New York

CRC Press is an imprint of the
Taylor & Francis Group, an **informa** business

A CHAPMAN & HALL BOOK

Taylor & Francis Group
6000 Broken Sound Parkway NW, Suite 300
Boca Raton, FL 33487-2742

First issued in paperback 2019

ISBN-13: 978-1-4822-6384-8 (hbk)
ISBN-13: 978-0-367-37730-4 (pbk)

Visit the Taylor & Francis Web site at
http://www.taylorandfrancis.com

and the CRC Press Web site at
http://www.crcpress.com

In memory of Donato Trigiante

Contents

Foreword

In the second half of September 2008, I [1] organized the first edition of the *Symposium on Recent Trends in the Numerical Solution of Differential Equations*, within the *6-th International Conference of Numerical Analysis and Applied Mathematics, ICNAAM 2008*, held in Kos, Greece, on September 16–20, 2008.[2] There were eleven speakers scheduled for the symposium: among them Felice Iavernaro, delivering a talk entitled "Conservative Boundary Value Methods for the solution of polynomial Hamiltonian systems". This information can still be retrieved at the URL of the symposium:

http://web.math.unifi.it/users/brugnano/ICNAAM2008/

Even though the talk was only 25 mins long (this was the time-slot for each talk in the symposium), I was impressed by the elegant and simple idea behind the methods presented by Felice. I approached him, just after the talk, and remember saying to him: – *Felice, I have not understood all the details about the methods, but the idea you have outlined is an "idea"* – meaning that the important ideas, in Mathematics, are relatively few.

This was the first time I heard about the derivation of conservative numerical methods, for Hamiltonian problems, obtained by using the so called *discrete line integral approach*, which was the name that Felice and his co-author (Brigida Pace, a former PhD student of Felice) gave to this methodological framework.[3] Felice answered to me: *I am glad that you liked my presentation,* – then continuing – *maybe we can collaborate on this topic.*

This was the beginning of our research collaboration, whose main achievements are covered in this book. The subsequent year, in the *Second Symposium on Recent Trends in the Numerical Solution of Differential Equations*, within the *7-th International Conference of Numerical Analysis and Applied Mathematics, ICNAAM 2009*, held in Rethymno, Greece, on September 18–22, 2009, the basic framework of the methods was presented in two talks of ours, still available at the URL of the symposium:

http://web.math.unifi.it/users/brugnano/ICNAAM2009/

---

[1] Luigi Brugnano.

[2] This is a conference series organized by Theodore Simos.

[3] In the previous edition of the same conference series (the *5-th International Conference of Numerical Analysis and Applied Mathematics, ICNAAM 2007*, held in Corfù, Greece, on September 16–20, 2007), Felice Iavernaro presented the first instance of such methods, named *s-stage trapezoidal methods*, which were generalized in his 2008 talk.

Actually, this research initially involved also our former advisor, professor Donato Trigiante, who passed away on September 18, 2011. This book is dedicated to his memory. We shall try, at our best, to follow his clear line of reasoning, based on the simplest possible mathematical arguments, in accordance with his preferred *motto* (slighly adapted from the Okham's Razor): "*Frustra fit per plura quod potest per pauciora*".[4]

We are sincerely grateful to Theodore Simos, the chairman of the ICNAAM Conference series, for having provided the venue for this project to start and consolidate, as well as for his support (also economical) for attending some of the ICNAAM editions.

We would like to express our sincere gratitude to all the people that have supported us during the preparation of the manuscript. Among them, we mention Lidia Aceto, Pierluigi Amodio, Gianluca Frasca Caccia, Cecilia Magherini, Francesca Mazzia, Juan Montijano, and Luis Rández, for reading parts of the manuscript and for providing us with valuable remarks and comments.

Last, but not least, we wish to thank our families, for having allowed us to devote a significant part of our free time to this project.

---

[4]I.e., "It is pointless to use more involved arguments, when simpler ones suffice."

Preface

> *. . . What we need is imagination.*
>
> Richard Feynman

This book deals with the numerical solution of differential problems within the framework of *Geometric Integration*, a branch of numerical analysis which aims to devise numerical methods able to reproduce, in the discrete solution, relevant geometric properties of the continuous vector field. Among them, a paramount role is played by the so called *constants of motion*, which are physical quantities that are conserved along the solution trajectories of a large set of differential systems, named *Conservative Problems*. In particular, the major emphasis will be on Hamiltonian systems, though more general problems will be also considered.

For canonical Hamiltonian problems, the most important constant of motion is the Hamiltonian function itself, which is often referred to as *the energy* of the system. For this reason, methods which are able to conserve the Hamiltonian are usually named *energy-conserving*. This book is meant to be a thorough, though concise, introduction to energy-conserving Runge-Kutta methods. The key tool exploited to devise these methods is what we have called *discrete line integral*: roughly speaking, one imposes energy conservation by requiring that a discrete counterpart of a line integral vanish along the numerical solution regarded as a path in the phase space.

This basic tool may be easily adapted to handle the conservation of multiple invariants. With the term *Line Integral Methods*, we collect all the numerical methods whose definition relies on the use of the line integral.

The material is arranged in order to provide, to the best of our ability, a quite friendly walk across the subject, yet still giving enough details to allow a concrete use of the methods, with a number of examples of applications. To this end, related Matlab software is downloadable at the homepage of the book (see Section A.2). On the other hand, the material is meant to be self-contained: only a basic knowledge about numerical quadrature and Runge-Kutta methods is assumed.

The book is organized in six chapters and one appendix:

1. Chapter 1 contains a primer on line integral methods, and provides a basic introduction to Hamiltonian problems and symplectic methods.

2. Chapter 2 contains the description of a number of Hamiltonian problems, which are representative of a variety of applications. Most of the problems will be later used as workbench for testing the methods.

3. Chapter 3 contains the core of the theoretical results concerning the main instance of Line Integral Methods, i.e., the class of energy-conserving Runge-Kutta methods, also named Hamiltonian Boundary Value Methods (HBVMs), specifically devised for the Hamiltonian case.

4. Chapter 4 addresses, in great detail, the issue of the actual implementation of HBVMs, which is paramount, in order to recover in the numerical solution what expected from the theory. The problems presented in Chapter 2 are used to assess the methods.

5. Chapter 5 deals with the application of HBVMs to handle the numerical solution of Hamiltonian partial differential equations, when the space-variable belongs to a finite domain and appropriate boundary conditions are specified.

6. Chapter 6 sketches a number of generalizations of the basic energy-conserving methods: the extension to the case of multiple invariants; the case of general conservative problems; the numerical solution of Hamiltonian boundary value problems.

7. Appendix A, at last, contains some background material concerning Legendre polynomials, along with a brief description of the Matlab software implementing the methods, which has been made available on the internet.

Firenze and Bari, April 2015.

Luigi Brugnano and Felice Iavernaro.

Symbol Description

$\mathbb{R}, \mathbb{C}$ denote the set of real and complex numbers, respectively.

$\mathbb{R}^n$ denotes the set of real (column) vectors of length n.

$\mathbb{R}^{m \times n}$ denotes the set of real $m \times n$ matrices.

i denotes the imaginary unit (i.e., $i^2 = -1$).

$\bar{z}$ denotes the complex-conjugate of z.

$\text{Re}(z)$ denotes the real part of z.

$\text{Im}(z)$ denotes the imaginary part of z.

I_n, I denotes the $n \times n$ identity matrix. If the subscript is not specified, the dimension of I can be deduced from the context.

$O_{m \times n}, O$ denotes the $m \times n$ zero matrix. If the subscripts are not specified, the dimensions of O can be deduced from the context.

$\mathbf{1}_n, \mathbf{1}$ denotes the unit vector of length n. If the subscript is not specified, the dimensions of $\mathbf{1}$ can be deduced from the context.

$\mathbf{0}_n, \mathbf{0}$ denotes the zero vector of length n. If the subscript is not specified, the dimensions of $\mathbf{0}$ can be deduced from the context.

$\sigma(A)$ denotes the spectrum of matrix A.

Π_s denotes the set of complex polynomials of degree at most s.

$A^\top, v^\top$ denotes the transpose matrix of A or vector v.

∇f if $f : \mathbb{R}^m \to \mathbb{R}$, $\nabla f \in \mathbb{R}^m$ denotes the gradient of f; if $f : \mathbb{R}^m \to \mathbb{R}^\nu$, then $\nabla f = \left[\frac{\partial f_j}{\partial x_i} \right] \in \mathbb{R}^{m \times \nu}$.

$f'(x)$ denotes the Jacobian of f: $f' = \nabla f^\top$.

$\nabla_q f$ denotes the vector of partial derivatives of f with respect to the specified variables.

$\dot{f}$ denotes the (time) derivative of f.

$\ddot{f}$ denotes the second (time) derivative of f.

$f^{(j)}$ denotes the j-th derivative of f.

u_x denotes the partial derivative of u w.r.t. x. Similarly, for higher-order partial derivatives.

$:=, \ =:$ the item on the side of the ":" is defined by that on the other side.

$\equiv$ the items at its sides coincide.

$\propto$ the functions at its sides are proportional each other.

C^p is the vector space of continuous functions with p continuous derivatives.

Chapter 1

A primer on line integral methods

In this chapter we outline the basic ideas providing the methodological frame-work for "line integral methods", along with its application to derive energy-conserving methods for Hamiltonian problems. The exposition of this chapter has been deliberately made as simple as possible, so that even the reader who has never confronted such material should easily familiarize with it. The main aim here is to present the foundation of the theory and to support the motivation for devising such methods by means of some simple illustrative examples.[1]

1.1 A general framework

In this book, we are concerned with the numerical solution of (autonomous) *conservative problems*: namely, problems in the form [2]

$$\dot{y} = f(y), \qquad t \geqslant 0, \qquad y(0) = y_0 \in D \subset \mathbb{R}^m, \tag{1.1}$$

for which there exists a (smooth) function

$$L : \mathbb{R}^m \to \mathbb{R}^\nu,$$

such that L remains constant when evaluated along the solution $y(t)$ of (1.1), for any choice of the initial condition y_0:

$$L(y(t)) \equiv L(y_0), \qquad \forall\, t \geqslant 0 \ \text{ and } \ y_0 \in D. \tag{1.2}$$

In such a case, we say that problem (1.1) admits ν *constants of motion*, or *first integrals* (the components of the vector-valued function L). For the sake of

---

[1]We have also included an overview of Hamiltonian systems and introduced a few geometrical integrators for their correct simulation. The manuscript is not intended to provide an in-depth review of the numerical treatment of conservative problems. Several excellent monographs and expository papers on the subject are available: see for example the books by Feng and Quin [87], Hairer, Lubich and Wanner [105], Leimkulher and Reich [133], Sanz-Serna and Calvo [153], Stuart and Humphries [161], and Suris [163].

[2]For sake of simplicity, we shall set the initial condition at $t_0 = 0$. Moreover, we shall assume that the solution of (1.1) exists on the whole half-line $t \geqslant 0$, which is the case of greatest interest.

simplicity, we shall at first consider the case of a *single* invariant (i.e., $\nu = 1$), whereas, the case of *multiple* invariants (i.e., $\nu \geqslant 2$) will be considered later. Furthermore, we shall also assume that both f and L can be expanded in Taylor series.

Differentiating (1.2) with respect to the time t yields

$$0 = \frac{\mathrm{d}}{\mathrm{d}t} L(y(t)) = \nabla L(y(t))^\top \dot{y}(t) = \nabla L(y(t))^\top f(y(t)),$$

thus a necessary and sufficient condition for L to be an invariant is

$$\nabla L(y)^\top f(y) = 0, \qquad \forall y \in D. \tag{1.3}$$

This condition, though characterizing the conservation property at a continuous level, turns out to be no longer necessary if we consider a *discrete time dynamics*. Specifically, let us relax condition (1.2) by requiring

$$L(y(h)) = L(y(0)), \tag{1.4}$$

for a given *stepsize* $h > 0$. From the Fundamental Theorem of Calculus, it then follows that an equivalent requirement for (1.4) to hold is the vanishing of the corresponding *line integral*,[3] i.e.,

$$L(y(h)) - L(y(0)) = \int_0^h \frac{\mathrm{d}}{\mathrm{d}t} L(y(t)) \mathrm{d}t = \int_0^h \nabla L(y(t))^\top \dot{y}(t) \mathrm{d}t = 0. \tag{1.5}$$

Though (1.3) clearly implies (1.5), in general the converse is not true, the latter condition being a discrete analogue of the former. In other words, there exist other functions $\sigma(t) \in \mathbb{R}^m$, different from the solution $y(t)$, that make the line integral $\int_0^h \nabla L(\sigma(t))^\top \dot{\sigma}(t) \mathrm{d}t$ vanish. Since our final goal is to construct numerical methods that are able to reproduce in the discrete setting the conservation properties of the continuous problem, the previous argument suggests that, in devising such methods, one can focus on criteria for imposing the less stringent condition (1.5).

With this simple premise, *Line Integral Methods* are numerical methods for solving problem (1.1), locally on the interval $[0, h]$, devised by determining a suitable approximation to $y(t)$, say $\sigma(t)$, which satisfies the following requirements:

$$\sigma(0) = y_0 \equiv y(0), \tag{1.6}$$

$$\sigma(h) =: y_1 \approx y(h), \tag{1.7}$$

$$\int_0^h \nabla L(\sigma(t))^\top \dot{\sigma}(t) \mathrm{d}t = 0. \tag{1.8}$$

---

[3]Given a vector field $F : D \subset \mathbb{R}^m \to \mathbb{R}^m$ and a smooth parametric curve $\sigma : [a, b] \to \mathbb{R}^m$, the line integral of F along the curve $\sigma(t)$ is $\int_\sigma F(y) \cdot \mathrm{d}\sigma = \int_a^b F(\sigma(t)) \cdot \dot{\sigma}(t) \mathrm{d}t$, where $\cdot$ denotes the dot product: $F(\sigma(t)) \cdot \dot{\sigma}(t) = F(\sigma(t))^\top \dot{\sigma}(t)$.

Notice that (1.6)–(1.8) imply $L(y_1) = L(y_0)$. The approximation $\sigma(t)$ may be interpreted as a curve in the phase space along which the line integral of the vector field $\nabla L(y)$ is evaluated. This curve acts as the unknown of the problem and, as we will see in the sequel, its determination will emerge from making the line integral vanish, according to equation (1.8).

The whole procedure is then repeated starting from $t_1 = h$, using the new approximation y_1 as the initial value, and so on. In so doing, one computes a discrete approximation to the solution,

$$y_n \approx y(t_n), \qquad t_n = nh, \qquad n = 0, 1, \ldots,$$

such that

$$L(y_n) = L(y_0), \qquad \forall n = 0, 1, \ldots, \tag{1.9}$$

that is, the invariant L is conserved along the (discrete) numerical solution. In other words, (1.9) is the discrete counterpart of (1.2), for the given stepsize h.

It should be noticed that requirements (1.6)–(1.8) are not yet sufficient, in general, to obtain a *practical* numerical method, due to the presence of an integral in (1.8). A quotation from Dahlquist and Björk reflects this issue:

> *As is well known, even many relatively simple integrals cannot be expressed in finite terms of elementary functions, and thus must be evaluated by numerical methods.*[4]

Consequently, the integral in equation (1.8), as well as all other possible integrals that might appear to fulfill it, will be approximated by means of a suitable quadrature formula with abscissae $0 \leqslant c_1 < \cdots < c_k \leqslant 1$, and corresponding weights $b_1, \ldots, b_k$. As a result, one obtains a *discrete line integral* approximating the continuous one

$$\int_0^h \nabla L(\sigma(x))^\top \dot{\sigma}(x) \mathrm{d}x \approx h \sum_{i=1}^k b_i \nabla L(\sigma(c_i h))^\top \dot{\sigma}(c_i h),$$

and, therefore, the original problem (1.6)–(1.8) is replaced by the following one:

$$u(0) = y_0 \equiv y(0), \tag{1.10}$$

$$u(h) =: y_1 \approx y(h), \tag{1.11}$$

$$\sum_{i=1}^k b_i \nabla L(u(c_i h))^\top \dot{u}(c_i h) = 0. \tag{1.12}$$

The use of a new curve $u(t)$ instead of $\sigma(t)$ reflects the fact that a solution

---

[4]G. Dahlquist and Å. Björk. *Numerical Methods in Scientific Computing, Volume I.* SIAM, Philadelphia, 2008, p. 521.

of this latter full-discretized problem will differ, in general, from the corresponding solution of the original problem but, as is easily argued, the error $\sigma(t) - u(t)$ will be of the same order as the error in the interpolatory quadrature formula. In particular, if the quadrature is exact, then $u(t)$ will match $\sigma(t)$.

As was observed above, the practical implications of handling problem (1.10)–(1.12) is that it will induce a method ready for implementation on a computer. On the other hand, from a theoretical viewpoint, it should also be noted that (1.10)–(1.12) define a sequence of problems depending on the positive index k. This sequence admits (1.6)–(1.8) as limit, which implies that $u(t) \to \sigma(t)$ as $k \to \infty$. We will see that increasing k does not affect too much the bulk of computational effort associated with the implementation of the method so, if necessary, one can choose k large enough that, in finite precision arithmetic, $u(t)$ and $\sigma(t)$ become undistinguishable. Modulo this assumption, which will be realized in practical computation, the two problems become equivalent.

The requirements (1.10)–(1.12) form the framework through which *(discrete) line integral methods* come to life. Depending on the particular shape of problem (1.1), all methods in this framework will be characterized by the choices of the approximation $\sigma(t)$ and of the quadrature $(c_i, b_i)_{i=1,\dots,k}$.

Remark

It turns to be advantageous to choose $\sigma(t)$ in a finite dimensional subspace of functions, since the problem to be solved will then take the form of a finite dimensional nonlinear system of equations. Throughout the manuscript, we will assume that $\sigma(t)$ is a polynomial curve but, in principle, other possibilities may also be considered. Moreover, for brevity, we shall sometimes use the acronym *LIM* in place of *Line Integral Method*.

1.2　Geometric integrators

From a geometric point of view, (1.2) means that the solution of problem (1.1) lies on the manifold

$$L(y) = const \equiv L(y(0)). \tag{1.13}$$

In the following, we shall always assume that the ν invariants are functionally independent, which means that all the considered y are *regular points* for the above constraints, that is, $\nabla L(y)$ has full (column) rank. Consequently, (1.13) identifies a $(m-\nu)$-dimensional manifold. Geometric properties such as (1.13) often play an important role in the mathematical modeling of a wide range of applications, ranging from the nano-scale of molecular dynamics to the macro-scale of celestial mechanics.

It has become customary to refer to numerical methods, able to reproduce *geometrical properties* of the continuous solution in the discrete approximation, as *geometric integrators*. It should be noted that the numerical solutions computed by traditional, general purpose codes often exhibit a wrong qualitative behavior when certain critical quantities are not conserved, so they turn out to be inappropriate when one is interested in reproducing the global, long-time behavior of the solutions of a conservative system in the phase space. In contrast with classical methods that focus on a control of the local discretization error, the main objective of a geometric integrator is to capture specific qualitative features of the problem at hand, providing a better reproduction of the topological properties of the solutions in the phase space.

The importance of geometric properties, like (1.13), stems from the fact that invariants of motion may represent meaningful physical properties of the underlying phenomenon. It is then important to reproduce them in the numerical simulation of the phenomenon itself. Conversely, the dynamics could be not reliably simulated.

In order to make this statement clear, let us consider the following very simple problem,

$$\dot q = p, \qquad \dot p = -q, \qquad q(0) = 1, \qquad p(0) = 0, \tag{1.14}$$

whose solution is easily seen to be given by

$$q(t) = \cos t, \qquad p(t) = -\sin t, \qquad t \geqslant 0. \tag{1.15}$$

Consequently, the function

$$L(q,p) = q^2 + p^2 \equiv \left\| \begin{pmatrix} q \\ p \end{pmatrix} \right\|_2^2 \tag{1.16}$$

assumes the constant value 1, along the solution of the problem, meaning that it lies on the unit circumference, in the phase-space (q,p).[5]

Approximating the solution of (1.14) by means of the explicit Euler method, with fixed stepsize h, yields:

$$q_{n+1} = q_n + hp_n, \qquad p_{n+1} = p_n - hq_n.$$

That is,

$$\begin{pmatrix} q_{n+1} \\ p_{n+1} \end{pmatrix} = \begin{pmatrix} 1 & h \\ -h & 1 \end{pmatrix} \begin{pmatrix} q_n \\ p_n \end{pmatrix},$$

---

[5]In general, a trajectory of (1.14) starting at $(q_0, p_0)^\top$ will lie on the circumference of radius $r = \|(q_0, p_0)^\top\|_2$. Equations (1.14) model an harmonic oscillator with total energy $E = \frac{1}{2}(q_0 + p_0)^2 = \frac{1}{2}L(q_0, p_0)$. That the solution lies on a circumference means that the energy remains constant during the motion.

from which one easily derives:

$$\begin{pmatrix} q_n \\ p_n \end{pmatrix} = \begin{pmatrix} 1 & h \\ -h & 1 \end{pmatrix}^n \begin{pmatrix} q_0 \\ p_0 \end{pmatrix} =: A_h{}^n \begin{pmatrix} 1 \\ 0 \end{pmatrix}, \qquad n = 0, 1, \ldots.$$

Since the eigenvalues of A_h are given by $1 \pm ih$, with i being the imaginary unit, one has that the discrete solution spirals outwards the unit circumference.

In a similar way, by using the implicit Euler method, one would obtain the discrete approximation

$$\begin{pmatrix} q_n \\ p_n \end{pmatrix} = \begin{pmatrix} 1 & -h \\ h & 1 \end{pmatrix}^{-n} \begin{pmatrix} q_0 \\ p_0 \end{pmatrix} = (A_h^\top)^{-n} \begin{pmatrix} 1 \\ 0 \end{pmatrix}, \qquad n = 0, 1, \ldots,$$

which now spirals inwards the unit circumference. Missing the important feature of keeping constant the Euclidean norm of the approximate solution, both the explicit and implicit Euler methods evidently are not geometric integrators for problem (1.14). On the other hand, the use of the trapezoidal rule would produce the sequence

$$\begin{aligned}
\begin{pmatrix} q_n \\ p_n \end{pmatrix} &= \left[\begin{pmatrix} 1 & -\frac{h}{2} \\ \frac{h}{2} & 1 \end{pmatrix}^{-1} \begin{pmatrix} 1 & \frac{h}{2} \\ -\frac{h}{2} & 1 \end{pmatrix} \right]^n \begin{pmatrix} q_0 \\ p_0 \end{pmatrix} \\
&= \left[A_{\frac{h}{2}}^{-\top} A_{\frac{h}{2}} \right]^n \begin{pmatrix} 1 \\ 0 \end{pmatrix} =: B_h{}^n \begin{pmatrix} 1 \\ 0 \end{pmatrix}, \qquad n = 0, 1, \ldots.
\end{aligned}$$

In such a case, since

$$A_{\frac{h}{2}}^{-\top} = \frac{1}{1 + \frac{h^2}{4}} A_{\frac{h}{2}},$$

the two matrices $A_{\frac{h}{2}}^{-\top}$ and $A_{\frac{h}{2}}$ commute, so that

$$B_h{}^\top B_h = \left(A_{\frac{h}{2}}^{-\top} A_{\frac{h}{2}} \right)^\top \left(A_{\frac{h}{2}}^{-\top} A_{\frac{h}{2}} \right) = A_{\frac{h}{2}}^\top A_{\frac{h}{2}}^{-1} A_{\frac{h}{2}} A_{\frac{h}{2}}^{-\top} = I.$$

Consequently, matrix B_h is orthogonal and, therefore:

$$\left\| \begin{pmatrix} q_n \\ p_n \end{pmatrix} \right\|_2 = \left\| \begin{pmatrix} q_0 \\ p_0 \end{pmatrix} \right\|_2 = 1, \qquad n = 0, 1, \ldots.$$

We then conclude that, for problem (1.14), the trapezoidal rule has the desired geometric property. Therefore, it may be considered a geometric integrator for the problem at hand. All the above facts are summarized by the plots on the left of Figure 1.1. The qualitative properties of the discrete solutions can also be checked by looking at the behavior of the invariant (1.16) computed along the numerical solution: in the right-picture of Figure 1.1 we see that the plots related to the two Euler methods exhibit a *drift* from the exact value which, instead, is correctly reproduced by the trapezoidal rule. Consequently, a (numerical) drift from the correct constant of motion usually means that the

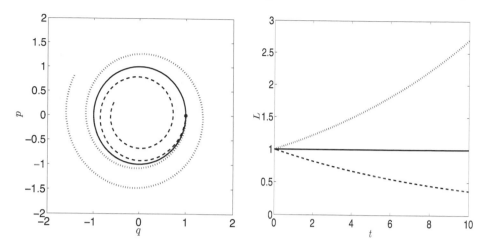

FIGURE 1.1: Left-plot: discrete approximations for problem (1.14), obtained by the explicit Euler method (dotted line), the implicit Euler method (dashed line), and the trapezoidal rule (solid line), by using a stepsize $h = 0.1$. The solid dot denotes the starting point of the trajectories. Right-plot: corresponding values of the invariant (1.16) along the discrete approximations, obtained by the explicit Euler method (dotted line), the implicit Euler method (dashed line), and the trapezoidal rule (solid line).

simulation is not correct. Geometric integrators are then expected to exhibit no drift in the invariant.

It is worth emphasizing that problem (1.14), though very simple, may be wrongly simulated numerically because its solution is only marginally (rather than asymptotically) stable. Consequently, the perturbations introduced by a "wrong" numerical method may cause its destruction, unless marginal stability is preserved. As a matter of fact, the two Euler methods transform the center configuration of the origin into a focus, either stable (implicit Euler) or unstable (explicit Euler); this will be the case, whichever the stepsize h used, so that a long-term simulation of (1.14) cannot be correctly done by using such methods. Conversely, the trapezoidal rule is able to preserve the correct stability pattern in the discrete solution, independently of the stepsize.

Though the problem we have solved is linear, the example above captures one important aspect of more complicated dynamical systems: the delicacy in reproducing the correct behavior in a neighborhood of marginally stable equilibrium points or periodic orbits. For a nonlinear and slightly more involved example, think about the two-body problem formed, for example, by the earth and a satellite. Depending on the value of energy, the orbit of the satellite could be elliptic, hyperbolic or parabolic. Assuming that the true solution be elliptic with energy E_0, making the value of energy associated with the numerical solution close enough to E_0 will prevent the satellite to depart from or fall towards the earth in a collision orbit as the time progresses.

Concerning the long-time behavior, the effect of a numerical method is similar to the introduction of a perturbing term in the original problem. If the dynamics takes place around a hyperbolic equilibrium point,[6] the theorem of stability in first approximation may be exploited to state whether the method preserves the character of that critical point.[7] A linear stability analysis is usually carried out to find out possible constrains on the stepsize h in order that the continuous and discrete solutions share the same asymptotic character. It turns out that methods with a nonempty absolute stability region are able to yield a correct simulated dynamics.

As an example, let us consider the following nonlinear perturbation of problem (1.14):

$$\begin{aligned}
\dot{q} &= p + \mu q(1 - q^2 - p^2), & q(0) &= 1, \\
\dot{p} &= -q + \mu p(1 - q^2 - p^2), & p(0) &= 0,
\end{aligned} \qquad (1.17)$$

with μ being a positive parameter, whose solution is still given by (1.15). The unit circumference is still an invariant set for (1.17), but is now asymptotically stable. Consequently, for any sufficiently small stepsize h, all of the above three methods will provide a qualitatively correct solution. As matter of fact, by setting $\mu = 1$, the discrete solution will approach a circumference of radius

- 1 for the trapezoidal rule;

- $1 + |O(h)|$ for the explicit Euler method;

- $1 - |O(h)|$ for the implicit Euler method.

Figure 1.2 confirms the above conclusions: the left-plot shows the numerical trajectories in the phase-space, obtained by using the same stepsize $h = 0.1$ previously considered for problem (1.14); the right-plot shows the numerical values of the invariant (1.16) computed along the corresponding numerical solutions for (1.17). As is clear, no drift now occurs.

This problem is less sensitive than the previous one, due to the more favorable stability properties of the solution set. Most of the methods and techniques employed in classical codes can successfully handle such situations. However, for conservative systems, a linear stability analysis does no longer suffice, and one needs much more sophisticated tools to ascertain the extent to which the numerical approximations can mimic the behavior of the exact solutions.

From the two examples above, one can argue that the correct approximation of the constants of motion somehow measures the reliability of the

---

[6] A hyperbolic critical point for $f(y)$ is a point y^* such that $f(y^*) = 0$ and $f'(y^*)$ has no eigenvalues with zero real part.

[7] The consistency conditions imply that the method preserves all critical points of the continuous problem.

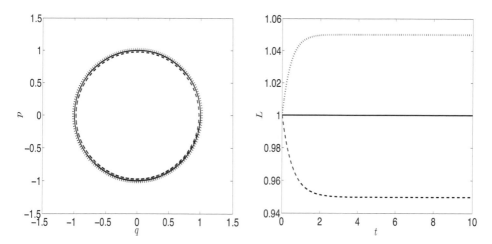

FIGURE 1.2: Left-plot: discrete approximations for problem (1.17), obtained by the explicit Euler method (dotted line), the implicit Euler method (dashed line), and the trapezoidal rule (solid line), by using a stepsize $h = 0.1$. Right-plot: corresponding values of the invariant (1.16) along the discrete approximations, obtained by the explicit Euler method (dotted line), the implicit Euler method (dashed line), and the trapezoidal rule (solid line).

underlying simulation. In particular, problem (1.14) is the simplest possible instance of a *Hamiltonian problem*: in the next section we introduce the general class of canonical Hamiltonian systems and discuss a few relevant properties of their solutions.

1.3 Hamiltonian problems

Let us consider a smooth scalar function $H(q, p)$, where $q, p \in \mathbb{R}^m$. The *canonical Hamiltonian problem* defined by H is

$$\dot{q} = \nabla_p H(q, p), \qquad \dot{p} = -\nabla_q H(q, p), \qquad (1.18)$$

where q is the vector of *generalized coordinates* and p is the vector of *conjugate momenta*. The entries q_i of the vector q uniquely specify the instantaneous configuration of the system: they might be, for example, Cartesian coordinates, polar coordinates, angles, or some mixture of these types of coordinates. Here, it is assumed that each of the q_i can vary independently, which means that the number of degrees of freedom is m.[8]

By setting

$$y = \begin{pmatrix} q \\ p \end{pmatrix} \in \mathbb{R}^{2m}, \qquad J = \begin{pmatrix} & I_m \\ -I_m & \end{pmatrix}, \tag{1.19}$$

problem (1.18) can also be written in a more compact form as

$$\dot{y} = J\nabla H(y). \tag{1.20}$$

The *Hamiltonian function H* is a conserved quantity. In fact, along a solution $y(t)$ of (1.20), one has:

$$\frac{\mathrm{d}}{\mathrm{d}t} H(y(t)) = \nabla H(y(t))^\top \dot{y}(t) = \nabla H(y(t))^\top J\nabla H(y(t)) = 0, \tag{1.21}$$

since matrix J (see (1.19)) is skew-symmetric. Consequently, for autonomous Hamiltonian systems, the Hamitonian function is *always* a constant of motion. More in general, we shall name Hamiltonian problem, any problem in the form (1.20), with matrix J multiplied by a constant scalar factor.[9]

From (1.3), we see that a scalar function $L : \mathbb{R}^{2m} \to \mathbb{R}$ is a first integral of (1.20) if and only if, for all y,

$$\nabla L(y)^\top J\nabla H(y) = 0. \tag{1.22}$$

This condition is often expressed in terms of *Poisson brackets*. Let us recall that, for any pair of regular functions $F, G : \mathbb{R}^{2m} \to \mathbb{R}$, their Poisson bracket $\{F, G\}$ is a function $\mathbb{R}^{2m} \to \mathbb{R}$ defined as

$$\{F, G\} \quad := \quad \nabla F^\top J\nabla G \quad \equiv \quad \nabla_q F^\top \nabla_p G - \nabla_p F^\top \nabla_q G.$$

Thus (1.22) is equivalent to $\{L, H\} = 0$. If two functions F and G satisfy the condition $\{F, G\} = 0$, they are said to be in involution.

A wide class of Hamiltonian systems are characterized by a *separable* Hamiltonian, meaning that it can be written in the form

$$H(p, q) = T(p) + U(q).$$

In such a case, problem (1.18) is said to be *separable*, and reduces to the two simpler equations

$$\dot{q} = \nabla T(p), \qquad \dot{p} = -\nabla U(q). \tag{1.23}$$

Several methods have been tailored to handle this special but important case.

Hamiltonian problems often occur in the modelization of isolated mechanical systems, for which H assumes the physical meaning of the total energy. For this reason, the Hamiltonian function is usually also referred to as the *energy*. For the same reason, geometric numerical integrators, that are able to conserve the Hamiltonian, are called *energy-conserving methods*.

---

[8]The number of degrees of freedom is defined as the minimum number of generalized coordinates necessary to define the configuration of the system. Thus, we are assuming that there are no mathematical constraints.

[9]Notice that (1.21) continues to hold, in such a case.

1.3.1 Symplecticity

Another important feature of Hamiltonian dynamical systems is that they possess a *symplectic* structure. To introduce this property we need a couple of ingredients:

- The *flow of the system*: it is the map acting on the phase space $\mathbb{R}^{2m}$ as

$$\phi_t : y_0 \in \mathbb{R}^{2m} \longrightarrow y(t) \in \mathbb{R}^{2m},$$

where $y(t)$ is the solution at time t of (1.20) originating from the initial condition y_0. Differentiating both sides of (1.20) with respect to y_0 and observing that

$$\frac{\partial y(t)}{\partial y_0} = \frac{\partial \phi_t(y_0)}{\partial y_0} \equiv \phi_t'(y_0),$$

we see that the Jacobian matrix of the flow ϕ_t is the solution of the variational equation associated with (1.20), namely

$$\frac{\mathrm{d}}{\mathrm{d}t} A(t) = J\nabla^2 H(y(t))A(t), \qquad A(0) = I, \tag{1.24}$$

where $\nabla^2 H(y)$ is the Hessian matrix of $H(y)$.

- The definition of a *symplectic transformation*: a map $u : (q, p) \in \mathbb{R}^{2m} \mapsto u(q, p) \in \mathbb{R}^{2m}$ is said *symplectic* if its Jacobian matrix $u'(q, p) \in \mathbb{R}^{2m \times 2m}$ is a symplectic matrix, that is

$$u'(q, p)^\top J u'(q, p) = J, \quad \text{for all } q, p \in \mathbb{R}^m.$$

With these premises, it is not difficult to show that, under regularity assumptions on $H(q, p)$, the flow associated with a Hamiltonian system is symplectic. Indeed, by setting

$$A(t) = \frac{\partial \phi_t}{\partial y_0},$$

and considering (1.24), one has

$$\frac{\mathrm{d}}{\mathrm{d}t} \left(A(t)^\top J A(t) \right) = \left(\frac{\mathrm{d}}{\mathrm{d}t} A(t) \right)^\top J A(t) + A(t)^\top J \left(\frac{\mathrm{d}}{\mathrm{d}t} A(t) \right)$$

$$A(t)^\top \nabla^2 H(y(t)) \underbrace{J^\top J}_{=I} A(t) + A(t)^\top \underbrace{JJ}_{=-I} \nabla^2 H(y(t))A(t) = 0.$$

Therefore

$$A(t)^\top J A(t) \equiv A(0)^\top J A(0) = J, \qquad \forall t \geqslant 0.$$

The converse of the above property is also true: if the flow associated with a dynamical system $\dot{y} = f(y)$ defined on $\mathbb{R}^{2m}$ is symplectic, then, at least locally in a neighbourhood of y_0, one has $f(y) = J\nabla H(y)$ for a suitable scalar function $H(y)$.

Symplecticity has relevant implications on the dynamics of Hamiltonian systems. Among the most important are:

(i) *Canonical transformations.* A change of variables $y = \psi(z)$ is *canonical*, that is, it preserves the structure of (1.20), with $F(z) = H(\psi(z))$ being the new Hamiltonian function, if and only if it is symplectic.[10] To see this, we observe that, after the change of variables, the new Hamiltonian system takes the form

$$\dot{z} = J\nabla F(z). \qquad (1.25)$$

Inserting $y = \psi(z)$ in (1.20) yields

$$\dot{z} = \psi'(z)^{-1}J\nabla H(\psi(z)).$$

A comparison with (1.25) yields

$$\psi'(z)^{-1}J\nabla H(\psi(z)) \;=\; J\nabla F(z) \;=\; J\psi'(z)^{\top}\nabla H(\psi(z)),$$

and, since $H(y)$ is arbitrary, we get $\psi'(z)^{-1}J = J\psi'(z)^{\top}$ and, because $J^{\top} = -J = J^{-1}$, one arrives at

$$\psi'(z)^{\top}J\psi'(z) = J.$$

Canonical transformations were known from Jacobi and used to recast (1.20) in simpler form. Analogously, it is straightforward to check that a change of variables $y = \psi(z)$ preserves the Poisson bracket if and only if it is symplectic:

$$\{F \circ \psi, G \circ \psi\} \;=\; \{F, G\} \circ \psi \qquad \Longleftrightarrow \qquad \psi'(z)^{\top}J\psi'(z) = J.$$

(ii) *Volume preservation.* The flow ϕ_t of a Hamiltonian system is volume preserving in the phase space. Recall that if V is a (suitable) domain of $\mathbb{R}^{2m}$, we have:

$$\mathrm{vol}(V) = \int_V dy \;\Rightarrow\; \mathrm{vol}(\phi_t(V)) = \int_{\phi_t(V)} dy = \int_V \left|\det \frac{\partial\phi_t(y)}{\partial y}\right| dy.$$

However, since $\frac{\partial\phi_t(y)}{\partial y} \equiv A(t)$ is a symplectic matrix, from $A(t)^{\top}JA(t) = J$ it follows that $\det(A(t))^2 = 1$ for any t and, hence, $\mathrm{vol}(\phi_t(V)) = \mathrm{vol}(V)$. In general, volume preservation is a characteristic feature of divergence-free vector fields. Recall that the divergence of a vector field $f : \mathbb{R}^n \to \mathbb{R}^n$ is the trace of its Jacobian matrix:

$$\mathrm{div}\, f(y) = \frac{\partial f_1}{\partial y_1} + \frac{\partial f_2}{\partial y_2} + \cdots + \frac{\partial f_n}{\partial y_n},$$

---

[10]ψ is assumed to be a local diffeomorphism.

so that f is divergence-free if

$$\operatorname{div} f(y) = 0, \qquad \forall y.$$

The vector field $J\nabla H$ associated with a Hamiltonian system has zero divergence. In fact, since

$$J\nabla H = \left(\frac{\partial H}{\partial p_1}, \ldots, \frac{\partial H}{\partial p_m}, -\frac{\partial H}{\partial q_1}, \ldots, -\frac{\partial H}{\partial q_m} \right)^{\top}$$

one has

$$\operatorname{div} J\nabla H = \frac{\partial^2 H}{\partial q_1 \partial p_1} + \cdots + \frac{\partial^2 H}{\partial q_m \partial p_m} - \frac{\partial^2 H}{\partial p_1 \partial q_1} - \cdots - \frac{\partial^2 H}{\partial p_m \partial q_m} = 0.$$

An important consequence of the previous property is Liouville theorem, which states that the flow ϕ_t associated with a divergence-free vector field $f : \mathbb{R}^n \to \mathbb{R}^n$ is volume preserving.

1.3.2   Integrable and nearly-integrable Hamiltonian systems

Coming back to item (i) above, if a symplectic transformation $(q,p) = \psi(\theta, I)$ exists such that the Hamiltonian function in the new variables does not depend on θ, that is $H(\psi(\theta, I)) = F(I)$, then the equations of motion can be solved analytically. In fact, after the canonical change of variables, system (1.18) becomes

$$\dot{\theta} = \omega(I) := F'(I), \qquad \dot{I} = 0, \qquad (1.26)$$

whose solutions are obtained by simple quadrature,

$$I(t) = I_0, \qquad \theta(t) = \theta_0 + \omega(I_0)t, \qquad (1.27)$$

where $(\theta_0, I_0) := \psi^{-1}(q_0, p_0)$ denotes the initial state.

As a simple example of one such transformation, consider again the harmonic oscillator (1.14) that now we write in a more general form:

$$\dot{q} = p, \qquad \dot{p} = -\omega^2 q, \qquad (1.28)$$

with energy function $H(q,p) = \frac{1}{2}\omega^2 q^2 + \frac{1}{2}p^2$. The general solution is

$$q(t) = A\cos(\omega t + \varphi), \qquad p(t) = -A\omega\sin(\omega t + \varphi),$$

where the constants A and φ are determined by imposing the initial conditions, and the motion takes place on the ellipse $H(q,p) = H(q_0, p_0)$. A direct computation shows that the change of variables $(q,p) = \psi(\theta, I)$ defined as

$$q = \sqrt{\frac{2I}{\omega}}\sin\theta, \qquad p = \sqrt{2\omega I}\cos\theta, \qquad (1.29)$$

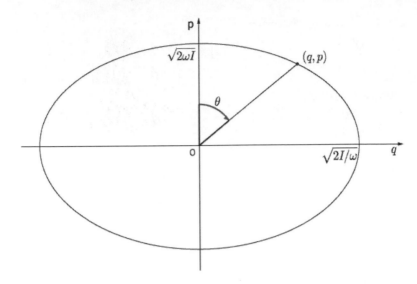

FIGURE 1.3: Canonical transformation from (q, p) to action-angle variables (θ, I) for the harmonic oscillator (1.28).

is symplectic, thus the new system is Hamiltonian with energy function given by:

$$F(\theta, I) = H(\psi(\theta, I)) = \omega I,$$

and, according to (1.27), its solution is

$$I(t) = I_0, \qquad \theta(t) = \theta_0 + \omega t.$$

By inverting the transformation (1.29), we find that

$$\frac{q^2}{\left(\frac{2I}{\omega}\right)} + \frac{p^2}{2\omega I} = 1, \tag{1.30}$$

thus, I_0 defines the ellipse while θ_0 specifies the initial anomaly. The coordinates I and θ are referred to as *action-angle* variables of the system. We observe that the action $I = I(q, p)$, as defined in (1.30), identifies a constant of motion representing a 1-torus, while θ is an angle coordinate on the torus and advances uniformly in time. Since two states (θ_1, I_0) and (θ_2, I_0) such that $\theta_2 - \theta_1$ is a multiple of 2π identify the same point in the phase space, we may assume $\theta \in \mathbb{T}^1 := \mathbb{R} \bmod 2\pi$.

In general, regarding the components θ_i of the state vector θ in (1.26)-(1.27) as angles, we can conveniently visualize the dynamics on the m-dimensional torus

$$\mathbb{T}^m := \mathbb{R}^m / 2\pi \mathbb{Z}^m \equiv \left\{ (\theta_1 \bmod 2\pi, \theta_2 \bmod 2\pi, \dots, \theta_m \bmod 2\pi)^\top, \theta_i \in \mathbb{R} \right\}.$$

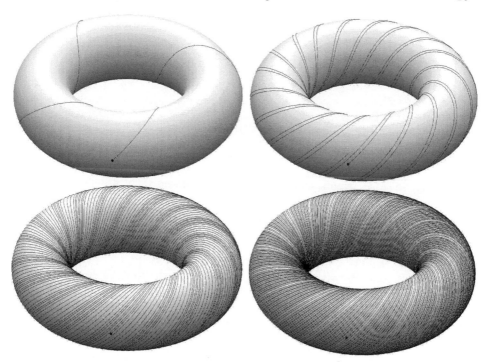

FIGURE 1.4: Dynamics on a 2-torus $\mathbb{T}^2$ of system (1.26) with $m = 2$. Since the two frequencies ω_1 and ω_2 have been chosen rationally independent, the orbit originating at the thick dot propagates on the 2-torus and covers it densely as the time progresses.

All vectors θ whose components differ by a multiple of 2π are identified by a single point on the torus. The vector $\omega(I_0)$ is the *frequency vector*. Its components, the frequencies ω_i, are said to be *rationally independent* (or *nonresonant*) if

$$\langle \omega, k \rangle := \sum_{i=1}^{m} k_i \omega_i \neq 0, \qquad \text{for all} \qquad k = (k_1, \ldots, k_m) \in \mathbb{Z}^m \backslash \{0\}. \quad (1.31)$$

In such a case, the corresponding solution (1.27) is quasi-periodic and thus covers densely the torus $\mathbb{T}^m$.

Liouville-Arnold theorem provides sufficient conditions for a canonical transformation to action-angle variables to exist. Consider the Hamiltonian system (1.18) and assume that it admits m smooth first integrals $L_1 \equiv H$, L_2, $\ldots, L_m$, which are pairwise in involution: $\{L_i, L_j\} = 0$ for all $i, j = 1, \ldots, m$. Fix a level set

$$M_c := \{(q, p) \in \mathbb{R}^{2m} | L_1(q, p) = c_1, \ldots, L_m(q, p) = c_m\},$$

for $c = (c_1, \ldots, c_m)^\top \in \mathbb{R}^m$, and assume that M_c is regular (the m first integrals L_i are independent at M_c):

$$\text{rank}\left(\nabla_{(q,p)} L(q,p)\right) = m, \qquad \text{for all} \quad (q,p) \in M_c,$$

with $L(q,p) = (L_1(q,p), \ldots, L_m(q,p))^\top$. Consequently, M_c is a m-dimensional submanifold of $\mathbb{R}^{2m}$ and we further assume that it is compact and connected. Then, the following properties hold true:

(i) M_c is diffeomorphic to the m-dimensional torus $\mathbb{T}^m$.[11]

(ii) In a suitable neighborhood of M_c there exist action-angle variables, that is, there exists a canonical transformation $(q,p) = \psi(\theta, I)$, defined for $\theta \in \mathbb{T}^m$ and $I \in B \subset \mathbb{R}^m$ such that the new Hamiltonian $F := H(\psi(\theta, I))$ only depends on the actions: $F = F(I)$.

Hamiltonian systems satisfying the assumptions of Liouville-Arnold theorem are called *completely integrable* and may be cast in the form (1.26).

The behavior of the solutions of systems such as (1.26) is completely understood, so these systems were meant to be the very first step towards the comprehension of more complicated problems. For example, in celestial mechanics, the elliptical motion of two bodies interacting through their mutual gravitational forces may be explained by equations in the form (1.26), whereas its simplest extension to a three-body configuration may not. However, in many cases of interest, a multi-body problem may be regarded as a perturbation of a set of two-body systems and, after Newton, astronomers used the solution of these independent two-body problems as starting approximations which then were refined by an iterative technique that, at each step, added a correction to the previously computed solution.[12]

The early efforts to understand the dynamics of more complicated systems, by regarding them as perturbations of simpler and completely solvable configurations, led to the foundation of *perturbation theory* and dates back to the 18-th century. The problem was first formalized by Poincaré as the *Problème général del la Dynamique* and concerns the study of the equations

$$\dot{\theta} = \nabla_I F(\theta, I), \qquad \dot{I} = -\nabla_\theta F(\theta, I), \tag{1.32}$$

under the assumption that the Hamiltonian function $F(\theta, I)$ may be expanded

---

[11]Two manifolds M and N are diffeomorphic if there is a diffeomorphism from M to N, that is a bijective differentiable map $M \to N$ such that its inverse is differentiable as well.

[12]One such example is *the main problem of lunar theory* that studies the anomaly in the monthly evolution of the moon around the earth (*evection*), caused by the presence of the sun. A further example is the solar system: since the planets have small masses compared to that of the sun, at a first stage one can ignore all the planet-to-planet interactions and recast the problem as a set of independent sun-planet, two-body systems, where each planet describes elliptical orbits around the sun. The effect of the external planets in each subsystem introduces small perturbations which should be accounted for to improve the accuracy of the solutions of the original problem.

in terms of increasing powers of a small parameter μ, namely:

$$F(\theta, I) \quad = \quad F_0(I) + \mu F_1(\theta, I) + \mu^2 F_2(\theta, I) + \dots, \tag{1.33}$$

where F_0 depends only on the I variable, and F_1, F_2, $\dots$, are periodic functions of period 2π with respect to θ. The 2π-periodicity with respect to the variable θ of the functions F_i in (1.33) suggests that the dynamics may be conveniently visualized on $\mathbb{T}^m \times \mathbb{R}^m$.

We notice that, by choosing $\mu = 0$, the equations in (1.32) reduce to those in (1.26), thus the parameter μ is meant to account for the presence of weak forces that are introduced in a simpler, fully understood system. For $\mu \approx 0$, the perturbed Hamiltonian system (1.32) is "close" to the unperturbed one (1.26) and therefore is said to be *nearly integrable*. Set $\mu = 0$ and assume that, associated with an initial condition (θ_0, I_0) is a nonresonant frequency vector $\omega(I_0) := F_0'(I_0)$: the corresponding solution

$$\theta(t) \ = \ \theta_0 + \omega(I_0)t, \qquad I(t) \ = \ I_0,$$

will be dense on the torus $\mathbb{T}^m \times \{I_0\}$. The basic *stability problem* is whether the condition of quasi-periodicity persists under slight perturbations, that is, for $\mu \neq 0$ but small. In this event, the qualitative behavior of the trajectories in the phase space would be unaffected by small perturbation of the parameter μ in a suitable neighborhood of zero, and system (1.32) would be *structurally stable*.

Finding sufficient conditions for quasi-periodicity to persist under small perturbations of this parameter challenged many researchers since the times of Poincaré. Given a nearly integrable system such as (1.32), a main task in perturbation theory is to determine a canonical transformation ψ that conjugates the perturbed system to a completely integrable system. The classical approach to attack the problem was to obtain ψ by composing infinitely many *elementary* canonical maps ψ_s, according to an iteration procedure: at step s a new elementary canonical map ψ_s is introduced such that the transformed system becomes $O(\mu^{s+1})$-close to an unperturbed completely integrable system.

The main issue during the implementation of this procedure was the appearance of *small divisors*, namely expressions of the form $\langle \omega, k \rangle = \sum_{i=1}^{m} k_i \omega_i$ as denominators of the coefficients of the Fourier series employed during the construction of ψ_s. Under the assumption (1.31) (i.e., that the frequencies are rationally independent), the series is formally well defined. However, this does not prevent that arbitrarily many denominators may approach zero in such a way that the terms of the series grow without bound, thus causing the series to diverge.

It was not until sixty years after Poincaré statement of the general problem of dynamics that Kolmogorov delivered the first positive answer. He overcame the problem of small divisors by devising a Newton-like procedure to define a sequence of elementary canonical maps whose composition converges very rapidly to a map that conjugates the original perturbed system to its unperturbed counterpart. In 1954, Kolmogorov published a short paper in Russian containing a general outline of his approach. Some years later, Arnold and Moser provided a detailed proof (see [12], [3], and [109] for details).

The main ingredient to make the new approach work is to assume the frequency vector $\omega(I_0)$ not only nonresonant (see (1.31)) but sufficiently far from being resonant. Specifically, the frequencies ω_i must satisfy the following Diophantine condition:

$$\left| \sum_{i=1}^{m} k_i \omega_i \right|^{-1} \leqslant c_0 \left(\sum_{i=1}^{m} |k_i| \right)^{\tau}, \quad \forall k = (k_1, \ldots, k_m)^{\top} \in \mathbb{Z}^m \backslash \{0\}, \quad (1.34)$$

for some positive constants c_0 and τ. If, in addition, the Hessian matrix of F_0 is nonsingular when evaluated at I_0 (nondegeneracy condition), a $\mu_0 > 0$ exists such that for $|\mu| < \mu_0$, the perturbed system (1.32) is analytically conjugate to its linear part $\dot{\theta} = \omega(I)$, $\dot{I} = 0$. Consequently, the unperturbed torus $\mathbb{T}^m \times \{I_0\}$ is analytically deformed into a torus which is invariant for the perturbed system, whose solution remains quasi-periodic.

The framework stemming from the studies of Kolmogorov, Arnold, and Moser is referred to as KAM theory.[13]

1.4   Symplectic methods

The above properties, and the fact that symplecticity is a characterizing property of Hamiltonian systems, reinforce the search of symplectic methods for their numerical integration. A one-step method

$$y_1 = \Phi_h(y_0)$$

is per se a transformation of the phase space. Therefore the method is symplectic if Φ_h is a symplectic map, i.e., if

$$\frac{\partial \Phi_h(y_0)}{\partial y_0}^{\top} J \frac{\partial \Phi_h(y_0)}{\partial y_0} = J.$$

---

[13]A closely related problem in complex dynamics concerns the existence of centers. Given a complex analytic function $f(z) = e^{2\pi i \alpha} z + \sum_{k=2}^{\infty} a_k z^k$, with $\alpha \in (0, 1)$, a formidable question in complex dynamics was whether it was possible for f to be conjugate to its linear part. In such an event, the origin would be a center for the discrete dynamical system $z_n = f(z_{n-1})$, where z_0 is an initial point close to the origin. This problem was addressed by Pfeiffer in 1915 and challenged several prominent figures such as Julia, Fatou, and Cremer. It was finally solved by Siegel in 1942 (prior to Kolmogorov), who proved the existence of centers assuming a Diophantine condition analogous to (1.34). The *center problem* is a small divisors problem, and Siegel's solution played an inspirational role in the development of Moser's approach to solve the small divisors problem in the context of celestial mechanics [3].

An important consequence of symplecticity in Runge-Kutta methods is the conservation of all *quadratic first integrals* of a Hamiltonian system.[14]

Lasagni, and independently Sanz-Serna and Suris, found out an easy criterion characterizing symplecticity of a Runge-Kutta method with tableau

$$\frac{\mathbf{c} \quad | \quad A}{\quad | \quad \mathbf{b}^\top} \qquad (1.35)$$

where, as usual, $\mathbf{c} = (c_i) \in \mathbb{R}^s$ is the vector of the abscissae, $\mathbf{b} = (b_i) \in \mathbb{R}^s$ is the vector of the weights, and $A = (a_{ij}) \in \mathbb{R}^{s \times s}$ is the corresponding Butcher matrix.

Theorem 1.1 ([152]). *The Runge-Kutta method (1.35) is symplectic if and only if*

$$\Omega A + A^\top \Omega = \mathbf{b} \mathbf{b}^\top, \qquad where \qquad \Omega = \mathrm{diag}(\mathbf{b}). \qquad (1.36)$$

It is worth noticing that, while in the continuous setting energy conservation derives from the property of *symplecticity* of the flow, the same is not necessarily true in the discrete setting: in fact, a symplectic integrator only preserves quadratic Hamiltonian functions. Attempts to incorporate both features into a numerical method culminated in a non-existence result due to Ge and Marsden (1988):

> *If [the method] is symplectic, and conserved H exactly, then it is the time advance map for the exact Hamiltonian system up to a reparametrization of time.*

This statement refers to nonintegrable Hamiltonian systems, that is, systems that do not admit other independent first integrals different from the Hamiltonian function itself. In 2005, Chartier, Faou, and Murua extended the previous result to general Hamiltonian systems, under the assumption that the integrator is a B-series method:[15]

> *The only symplectic method (as B-series) that conserves the Hamiltonian for arbitrary $H(y)$ is the exact flow of the differential equation.*

Non-existence results such as the one stated above beg two fundamental questions:

1. To what extent are symplectic methods prone to reproducing the energy-preserving property? And conversely,

2. May energy-preserving methods mimic the geometric behavior of a symplectic method?

---

[14] A quadratic first integral takes the form $L(y) = y^\top C y$, with C a symmetric matrix.
[15] Runge-Kutta methods admit an expansion in terms of B-series.

We briefly comment on these two issues separately, and refer the reader to the sources listed in footnote 1 on page 1 for a thorough analysis.

1. Concerning the first item, under suitable regularity assumptions, a backward analysis approach shows that, when a symplectic method $y_1 = \Phi_h(y_0)$ of order p is used with a constant stepsize, the numerical solution formally satisfies a perturbed Hamiltonian problem that is expressed as a power series in the stepsize h. Specifically, for $t_n = nh$, the numerical solution is requested to satisfy $y_n = z(t_n)$, where $z(t)$ is the solution of the perturbed problem

$$\dot{z}(t) \;=\; J\nabla\bar{H}(z(t)),$$

$$\bar{H}(z) \;=\; H(z) + h^p H_p(z) + h^{p+1} H_{p+1}(z) + \dots.$$

Should this series converge, we would get a near conservation of energy over infinite time intervals:

$$H(y_n) = H(y_0) + O(h^p), \qquad \forall t_n \geqslant t_0. \tag{1.37}$$

Unfortunately, the convergence of the series is not guaranteed in general, so one has to truncate it and study the effect on the energy of the numerical solution y_n. In 1994, Benettin and Giorgilli showed that, if the numerical solution stays within a compact set and $h \leqslant h_0$, then

$$H(y_n) \;=\; H(y_0) + O(h^p) + O(t_n e^{-\frac{C}{h}}), \tag{1.38}$$

where the positive constant C depends only on the method and the Lipschitz constant of the problem. For h small enough, it turns out that the last term at the right hand side is negligible with respect to the $O(h^p)$ term over *exponentially long-times*: on such long intervals (1.38) becomes equivalent to (1.37).

Backward error analysis has proven a fundamental tool in the study of the long-time behavior of the solution generated by geometric integrators. For example, combining backward error analysis and KAM theory, Calvo and Hairer showed that symplectic methods applied to completely integrable systems display a linear (rather than quadratic) error growth when the frequencies at the initial value satisfy the Diophantine condition (1.34):

$$y(t) - y_n \;=\; O(th^p)$$
$$I(y_n) - I(y_0) \;=\; O(h^p),$$

for $t_n = nh < h^{-p}$. The long-time conservation of the action variables also extend to the symplectic integration of nearly integrable systems.

2. Concerning the second item listed above, the first step is to check whether the method $y_1 = \Phi_h(y_0)$ is conjugate to a symplectic method $y_1 = \Psi_h(y_0)$. This means that a global change of coordinates $\chi_h(y) = y + O(h^p)$

exists such that $\Phi_h = \chi_h \circ \Psi_h \circ \chi_h^{-1}$. In this event, the solution $\{y_n\}$ computed by the former method would satisfy

$$y_n = \Phi_h^n(y_0) = (\chi_h \circ \Psi_h \circ \chi_h^{-1})^n(y_0) = \chi_h \circ \Psi_h^n \circ \chi_h^{-1}(y_0).$$

Thus, symplectic conjugate methods inherit the long-time behavior of symplectic integrators.

Conjugate symplecticity is, in general, a too demanding goal. A less stringent condition is to check whether the method $y_1 = \Psi_h(y_0)$ is conjugate symplectic up to the order $q = p + r$, for some nonnegative integer r:

$$\Phi_h = \chi_h \circ \Psi_h \circ \chi_h^{-1}, \qquad \text{with} \quad \Psi_h'(y)^\top J \Psi_h'(y) = J + O(h^{p+r+1}). \quad (1.39)$$

A method with property (1.39), also called *pseudo-symplectic*, shares the same long-time behavior of symplectic methods on time intervals of length $O(h^{-r})$. For example, (1.38) would become

$$H(y_n) = H(y_0) + O(h^p) + O(th^{p+r}).$$

1.4.1 Symplectic Euler method

Among the easiest and oldest methods yielding a symplectic transformation are the following variants of the classical Euler methods, adapted to handle separately the q and p variables in system (1.18):

$$q_1 = q_0 + h\nabla_p H(q_0, p_1), \qquad p_1 = p_0 - h\nabla_q H(q_0, p_1), \quad (1.40)$$

and

$$q_1 = q_0 + h\nabla_p H(q_1, p_0), \qquad p_1 = p_0 - h\nabla_q H(q_1, p_0). \quad (1.41)$$

Both formulae have order 1 and may be viewed as intermediate between the classical explicit and implicit Euler methods, in that (1.40) is explicit in the q variable and implicit in the p variable and, conversely, (1.41) is implicit in the q variable and explicit in the p variable. Interestingly, for separable Hamiltonian functions (1.23) they reduce to the following explicit methods:

$$q_1 = q_0 + h\nabla T(p_1), \qquad p_1 = p_0 - h\nabla U(q_0), \quad (1.42)$$

and

$$q_1 = q_0 + h\nabla T(p_0), \qquad p_1 = p_0 - h\nabla U(q_1). \quad (1.43)$$

Let us check the symplecticity property considering, for brevity, formula (1.42). One gets:

$$\frac{\partial p_1}{\partial q_0} = -h\nabla^2 U(q_0), \qquad \frac{\partial p_1}{\partial p_0} = I,$$

$$\frac{\partial q_1}{\partial q_0} = I + h\nabla^2 T(p_1)\frac{\partial p_1}{\partial q_0} = I - h^2 \nabla^2 T(p_1)\nabla^2 U(q_0),$$

$$\frac{\partial q_1}{\partial p_0} = h\nabla^2 T(p_1)\frac{\partial p_1}{\partial p_0} = h\nabla^2 T(p_1).$$

Hence,

$$\frac{\partial}{\partial y_0}\Phi_h(y) \equiv \begin{pmatrix} \dfrac{\partial q_1}{\partial q_0} & \dfrac{\partial q_1}{\partial p_0} \\[2mm] \dfrac{\partial p_1}{\partial q_0} & \dfrac{\partial p_1}{\partial p_0} \end{pmatrix} = \begin{pmatrix} I - h^2 \nabla^2 T(p_1)\nabla^2 U(q_0) & h\nabla^2 T(p_1) \\[2mm] -h\nabla^2 U(q_0) & I \end{pmatrix},$$

and a straightforward computation proves that $(\frac{\partial}{\partial y}\Phi_h(y))^\top J(\frac{\partial}{\partial y}\Phi_h(y)) = J$.

1.4.2   Störmer-Verlet method

The Störmer-Verlet method is defined as the composition of the two symplectic Euler methods with halved stepsize. We derive the method under the assumption of a separable Hamiltonian function, which will be the form we need for the numerical experiments throughout the manuscript. Denoting $y_1 = \Phi_h^{(1)}(y_0)$ and $y_1 = \Phi_h^{(2)}(y_0)$ methods (1.42) and (1.43) respectively, the composition $y_1 = \Phi_h(y_0) := \Phi_{h/2}^{(2)} \circ \Phi_{h/2}^{(1)}(y_0)$ reads

$$p_{1/2} = p_0 - \frac{h}{2}\nabla U(q_0), \quad q_1 = q_0 + h\nabla T(p_{1/2}), \quad p_1 = p_{1/2} - \frac{h}{2}\nabla U(q_1). \quad (1.44)$$

As a composition of two symplectic methods, the Störmer-Verlet method remains symplectic. Furthermore, we notice that (1.41) is the adjoint method of (1.40), that is

$$\Phi_h^{(2)} = \left(\Phi_{-h}^{(1)}\right)^{-1}.$$

Consequently the composition gives rise to a symmetric, order two scheme. In the special case where $T(p) = \frac{1}{2}p^\top p$, that is for second-order problems $\ddot{q}(t) = -\nabla U(q(t))$, the scheme (1.44) becomes

$$\frac{q_2 - 2q_1 + q_0}{h^2} = -\nabla U(q_1), \qquad p_1 = \frac{q_2 - q_0}{2h},$$

which means that $\ddot{q}(t)$ at time $t = t_0 + h$ is approximated by means of the second-order central difference quotient defined by q_0, q_1 and q_2.

The idea of composing low-order symplectic methods, to derive higher-order ones, can be generalized. As an example, composing the Störmer-Verlet method with the symmetric stepsizes

$$\frac{h}{2 - r_2}, \quad \frac{-r_2 h}{2 - r_2}, \quad \frac{h}{2 - r_2}, \qquad r_2 = \sqrt[3]{2},$$

yields a fourth-order symplectic method. In turn, the composition of the latter method with the symmetric stepsizes

$$\frac{h}{2 - r_4}, \quad \frac{-r_4 h}{2 - r_4}, \quad \frac{h}{2 - r_4}, \qquad r_4 = \sqrt[5]{2},$$

yields a sixth-order symplectic method, and so forth. In general, having got an (even) order p symplectic method, its composistion with the symmetric stepsizes

$$\frac{h}{2 - r_p}, \quad \frac{-r_p h}{2 - r_p}, \quad \frac{h}{2 - r_p}, \qquad r_p = {}^{p+1}\sqrt{2},$$

provides us with a symplectic method of order $p + 2$. Clearly, the order 4 method requires 3 steps of the basic Störmer-Verlet method, the order 6 method 9 steps and, in general, the order p method will require $3^{(p-2)/2}$ applications of the basic method.

1.4.3 Gauss-Legendre methods

Gauss-Legendre formulae form a relevant class of symplectic methods. They belong to the class of collocation methods, which are introduced by means of the solution of the following problem.

Consider the initial value problem $\dot{y} = f(y)$, with $y(0) = y_0$. Given a set of s distinct abscissae $0 \leqslant c_1 < \cdots < c_s \leqslant 1$ and a stepsize $h > 0$, find a polynomial $v(t)$ satisfying the *collocation conditions*

$$\begin{aligned} v(0) &= y_0, \\ \dot{v}(c_i h) &= f(v(c_i h)), \quad i = 1, \ldots, s. \end{aligned} \tag{1.45}$$

The polynomial $v(t)$ satisfying (1.45) is then used to advance the solution, by posing $y_1 = v(h)$. It is well known and easy to demonstrate that the above problem defines an s-stage Runge-Kutta method with tableau

$$\begin{array}{c|c} c & A \\ \hline & b^\top \end{array} \qquad a_{ij} = \int_0^{c_i} \ell_j(\tau) \mathrm{d}\tau, \quad b_i = \int_0^1 \ell_i(\tau) \mathrm{d}\tau, \tag{1.46}$$

where, for $i = 1, \ldots, s$, $\ell_i(t)$ is the i-th Lagrange polynomial defined on the abscissae c_i. To see this, one expands the polynomial, of degree $s - 1$, $\dot{v}(ch)$ along the Lagrange basis $\{\ell_j(c)\}_{j=1,\ldots,s}$, defined on the abscissae c_i:

$$\dot{v}(ch) = \sum_{j=1}^{s} \ell_j(c) \eta_j, \tag{1.47}$$

where the unknown vectors η_j are to be determined by imposing conditions (1.45). Integrating both sides of (1.47) with respect to the variable c yields

$$v(ch) = y_0 + h \sum_{j=1}^{s} \int_0^c \ell_j(\tau) \mathrm{d}\tau \, \eta_j. \tag{1.48}$$

Set $Y_i = v(c_i h)$ and evaluate (1.47) at the abscissae c_i to get, by virtue of the collocation conditions (1.45),

$$\eta_i = \dot{v}(c_i h) = f(v(c_i h)) = f(Y_i).$$

On evaluating (1.48) at the abscissae c_i and at $c = 1$ we finally obtain, respectively,

$$Y_i = y_0 + h \sum_{j=1}^{s} \int_0^{c_i} \ell_j(\tau) \mathrm{d}\tau \, f(Y_j), \qquad i = 1, \ldots, s,$$

$$y_1 = y_0 + h \sum_{j=1}^{s} \int_0^{1} \ell_j(\tau) \mathrm{d}\tau \, f(Y_j),$$

that is method (1.46). Compared with general, fully implicit Runge-Kutta methods, the study of collocation methods turns to be much simplified. For example, they inherit the order of the underlying quadrature formula (c_i, b_i). The order is then maximized (i.e., it is $2s$) by choosing a Gauss-Legendre distribution of the nodes c_i on the interval $[0,1]$, which is defined as the set of roots of the s-th Legendre polynomial shifted on the interval $[0,1]$.

In general, the j-th degree Legendre polynomial, shifted on the interval $[0,1]$ and normalized with respect to the L^2 norm, reads

$$P_j(c) = \frac{\sqrt{2j+1}}{j!} \frac{\mathrm{d}^j}{\mathrm{d}c^j} \left[(c^2 - c)^j \right], \qquad j = 0, 1, \ldots. \tag{1.49}$$

The set of Legendre polynomials (1.49) form an orthonormal basis for the square-integrable function space $L^2([0,1])$:

$$\int_0^1 P_i(\tau) P_j(\tau) \mathrm{d}\tau = \delta_{ij} := \begin{cases} 0, & \text{if } i \neq j, \\ 1, & \text{if } i = j, \end{cases} \qquad \forall i, j = 0, 1, \ldots. \tag{1.50}$$

As we will see in Chapter 3, *Runge-Kutta line integral methods* are intimately related to Gauss methods, in that they may be interpreted as a generalization of these latter formulae. For this reason, it turns out to be very useful to derive an alternative (but equivalent) formulation of Gauss methods with the aid of the shifted and normalized Legendre polynomials defined above. Thus, we consider again the collocation conditions (1.45) but now expand the polynomial $\dot{v}(ch)$ along the Legendre (rather than the Lagrange) polynomial basis:

$$\dot{v}(ch) = \sum_{j=0}^{s-1} P_j(c) \gamma_j, \tag{1.51}$$

where, again, the unknown vectors γ_j are to be determined by imposing conditions (1.45). Integrating both sides of (1.51) with respect to the variable c yields

$$v(ch) = y_0 + h \sum_{j=0}^{s-1} \int_0^c P_j(\tau) \mathrm{d}\tau \, \gamma_j. \tag{1.52}$$

In terms of the coefficients γ_j appearing in (1.51), the collocation conditions (1.45) become (as above we set $Y_i = v(c_i h)$)

$$\sum_{j=0}^{s-1} P_j(c_i)\gamma_j = f(Y_i), \qquad i = 1,\ldots,s,$$

or, in matrix notation,

$$\mathcal{P}_s \otimes I\,\gamma = f(Y), \tag{1.53}$$

where

$$\mathcal{P}_s = \begin{pmatrix} P_0(c_1) & \cdots & P_{s-1}(c_1) \\ \vdots & & \vdots \\ P_0(c_s) & \cdots & P_{s-1}(c_s) \end{pmatrix}, \quad Y = \begin{pmatrix} Y_1 \\ \vdots \\ Y_s \end{pmatrix}, \quad f(Y) = \begin{pmatrix} f(Y_1) \\ \vdots \\ f(Y_s) \end{pmatrix}.$$

In turn, evaluating (1.52) at the abscissae c_i yields, in matrix form,

$$Y = \mathbf{1} \otimes y_0 + h\mathcal{I}_s \otimes I\,\gamma, \tag{1.54}$$

where

$$\mathcal{I}_s = \begin{pmatrix} \int_0^{c_1} P_0(x)\mathrm{d}x & \cdots & \int_0^{c_1} P_{s-1}(x)\mathrm{d}x \\ \vdots & & \vdots \\ \int_0^{c_s} P_0(x)\mathrm{d}x & \cdots & \int_0^{c_s} P_{s-1}(x)\mathrm{d}x \end{pmatrix}. \tag{1.55}$$

Matrices such as $\mathcal{P}_s$ and $\mathcal{I}_s$ will be considered again in Chapter 3, where some properties of interest will be analyzed. One such a property is the following relation between $\mathcal{P}_s$ and $\mathcal{I}_s$:[16]

$$\mathcal{I}_s = \mathcal{P}_s X_s, \tag{1.56}$$

with

$$X_s = \begin{pmatrix} \frac{1}{2} & -\xi_1 & & \\ \xi_1 & 0 & \ddots & \\ & \ddots & \ddots & -\xi_{s-1} \\ & & \xi_{s-1} & 0 \end{pmatrix}, \quad \xi_i = \frac{1}{2\sqrt{4i^2 - 1}}, \quad i = 1,\ldots,s-1.$$

Exploiting (1.56) and (1.53), we can recast (1.54) as

$$Y = \mathbf{1} \otimes y_0 + h\mathcal{P}_s X_s \mathcal{P}_s^{-1} \otimes I f(Y). \tag{1.57}$$

Equation (1.57) is equivalent to the one previously derived by means of the Lagrange polynomials. It was first introduced in 1981 by Hairer and Wanner

---

[16]The proof is reported in Lemma 3.6 on page 96.

who developed a powerful tool to analyze several classes of implicit Runge-Kutta methods. In that context, matrix $\mathcal{P}_s$ is referred to as W-transformation and, as the name suggests, is denoted by W.[17]

The use of the W-transformation to describe the coefficient matrix of a Gauss-Legendre method gives us the opportunity to easily derive the symplecticity property of these formulae by exploiting conditions (1.36). To this end, we need the following properties of Legendre polynomials and of matrix $\mathcal{P}_s$:

1. $\displaystyle\int_0^1 P_0(c)dc = 1, \qquad \int_0^1 P_j(c)dc = 0, \quad j = 1, 2, \ldots;$

2. $\mathcal{P}_s^\top \boldsymbol{b} = \boldsymbol{e}_1,$ with $\boldsymbol{e}_1 = (1, 0, \ldots, 0)^\top;$

3. $\mathcal{P}_s^{-1} = \mathcal{P}_s^\top \Omega.$

Property 1 follows from the orthogonality conditions (1.50) and considering that $P_0(c) = 1$. Property 2 follows from observing that the Gauss-Legendre quadrature formula with s nodes has degree of precision $2s - 1$:

$$(\mathcal{P}_s^\top \boldsymbol{b})_i = \sum_{j=1}^s b_j P_{i-1}(c_j) = \int_0^1 P_{i-1}(c)dc = \delta_{i1}, \qquad i = 1, \ldots, s.$$

Property 3 may be deduced by the same reasoning.[18] We are now in the right position to show that $A = \mathcal{P}_s X_s \mathcal{P}_s^{-1}$ satisfies conditions (1.36):

$$\begin{aligned}
\Omega A &+ A^\top \Omega \\
&= \Omega \mathcal{P}_s X_s \mathcal{P}_s^{-1} + \mathcal{P}_s^{-\top} X_s^\top \mathcal{P}_s^\top \Omega \;=\; \mathcal{P}_s^{-\top} X_s \mathcal{P}_s^{-1} + \mathcal{P}_s^{-\top} X_s^\top \mathcal{P}_s^{-1} \\
&= \mathcal{P}_s^{-\top} \left(X_s + X_s^\top \right) \mathcal{P}_s^{-1} \;=\; \mathcal{P}_s^{-\top} \left(\boldsymbol{e}_1 \boldsymbol{e}_1^\top \right) \mathcal{P}_s^{-1} \;=\; (\mathcal{P}_s^{-\top} \boldsymbol{e}_1)(\mathcal{P}_s^{-\top} \boldsymbol{e}_1)^\top \\
&= \boldsymbol{b}\boldsymbol{b}^\top.
\end{aligned}$$

Consequently, all s-stage Gauss-Legendre collocation methods are symplectic, for $s = 1, 2, \ldots$.

As an example, we report the Butcher tableau of the 1-stage (order 2) and 2-stage (order 4) Gauss methods:

$$
\begin{array}{c|c}
\frac{1}{2} & \frac{1}{2} \\
\hline
 & 1
\end{array}
\qquad
\begin{array}{c|cc}
\frac{1}{2} - \frac{\sqrt{3}}{6} & \frac{1}{4} & \frac{1}{4} - \frac{\sqrt{3}}{6} \\
\frac{1}{2} + \frac{\sqrt{3}}{6} & \frac{1}{4} + \frac{\sqrt{3}}{6} & \frac{1}{4} \\
\hline
 & \frac{1}{2} & \frac{1}{2}
\end{array}
$$

In particular, the former formula defines a discrete problem in the form

$$Y_1 = y_0 + \frac{h}{2} f(Y_1), \qquad y_1 = y_0 + h f(Y_1).$$

---

[17] When needed, we will adopt the name W-transformation, even though we will use the notation $\mathcal{P}$ instead of W.

[18] See Lemma 3.6 for details.

Multiplying the former equation by 2, and subtracting the latter one, gives $Y_1 = (y_0 + y_1)/2$, so the method can be also rewritten as

$$y_1 = y_0 + hf\left(\frac{y_0 + y_1}{2}\right).$$

In this equivalent form, it is mostly often referred to as *implicit mid-point method*.

1.5 *s*-Stage trapezoidal methods

It is for Hamiltonian problems that the corpus of the theory of line integral methods has been first developed, forming the core part of the present manuscript. The idea of exploiting the line integral fits the *modus operandi* characterizing the spirit of geometric integration: one considers a specific geometrical property of the continuous problem and tries to transfer it to the method. It is often the geometrical property of interest that suggests how to construct the method: this is the case with line integral methods. The interplay between the continuous and the discrete line integrals, in association with the differential problem and its discrete counterpart, is summarized by the following diagram:

$$\dot{y} = J\nabla H(y) \longrightarrow \int_\sigma \nabla H(y) \cdot dy$$

$$\uparrow ? \qquad\qquad\qquad \downarrow$$

$$y_1 = \Phi_h(y_0) \underset{?}{\longleftarrow} \sum_\sigma \nabla H \cdot \Delta y$$

where $\sum_\sigma \nabla H \cdot \Delta y$ is a shorthand for the discrete line integral. The arrows outline the path we follow to define and analyze a line integral method for Hamiltonian problems. Starting from a canonical Hamiltonian system:

1. one considers the line integral of the associated conservative vector field $\nabla H(y)$;

2. a quadrature formula, used to solve the integral numerically, leads to the discrete line integral;

3. a method is defined by imposing that the discrete line integral vanishes;

4. the resulting method is finally compared to the original differential equation to analyze the convergence and conservation properties.

The first two phases have been described in Section 1.1, while the question marks in the above diagram mean that the corresponding phases have not been discussed yet. In the remaining part of the present chapter we will describe, in a quite general setting, point 3, while point 4 will be thoroughly faced in Chapter 3. Surprisingly, it turns out that when the curve $\sigma(t)$ is a polynomial, and the integrals are computed by means of an interpolatory quadrature formula, the resulting energy-preserving integrators are Runge-Kutta methods characterized by a low-rank coefficient matrix and intimately related to the classical collocation methods.

In this section, we study in some detail the simplest family of line integral methods, named *s-stage trapezoidal methods*. Their name stems from the fact that they may be regarded as a generalization of the basic trapezoidal method. To begin with, let us consider again problem (1.14), which is defined by the quadratic Hamiltonian

$$H(q,p) = \frac{1}{2}p^2 + \frac{1}{2}q^2. \tag{1.58}$$

By using the notation defined in (1.19),

$$y = \begin{pmatrix} q \\ p \end{pmatrix}, \qquad J = \begin{pmatrix} & 1 \\ -1 & \end{pmatrix} = -J^\top,$$

we can rewrite the Hamiltonian (1.58) as $H(y) = \frac{1}{2}y^\top y$, and the system reads

$$y = J\nabla H(y) \equiv Jy, \qquad y(0) = y_0. \tag{1.59}$$

Given a stepsize $h > 0$, we wish to define the simplest polynomial $\sigma(t)$, approximating $y(t)$ in the interval $[0, h]$, and such that $\sigma(0) = y_0$. Recall that our final goal is to achieve an approximation $y_1 := \sigma(h)$ yielding the energy-conservation property $H(y_1) = H(y_0)$. For problem (1.59) this task may be easily accomplished and the flow of reasoning will be later exploited to tackle more involved problems.

The simplest polynomial curve one can think of is, of course, the segment joining y_0 to y_1:

$$\sigma(t) = y_0 + t\,\frac{y_1 - y_0}{h}, \qquad t \in [0, h]. \tag{1.60}$$

We express the variation of the energy at the end points of $\sigma(t)$ in terms of the line integral of $\nabla H(y) = y$ along the curve $\sigma(t)$:

$$H(y_1) - H(y_0) = H(\sigma(h)) - H(\sigma(0)) = \int_0^h \dot{\sigma}(t)^\top \nabla H(\sigma(t))\mathrm{d}t. \tag{1.61}$$

Setting $t = ch$, one has then:

$$H(y_1) - H(y_0) = (y_1 - y_0)^\top \int_0^1 (y_0 + c(y_1 - y_0))\mathrm{d}c.$$

For this linear problem, it turns out that the integrand function is just the degree one polynomial $\sigma(ch)$. Consequently, the trapezoidal rule may be used to solve exactly the integral and we get:

$$
\begin{aligned}
H(y_1) - H(y_0) &= (y_1 - y_0)^\top \frac{y_0 + y_1}{2} \\
&= (y_1 - y_0)^\top \frac{\nabla H(y_0) + \nabla H(y_1)}{2}.
\end{aligned} \tag{1.62}
$$

Formula (1.62) suggests that, in order to get energy conservation, the numerical method should advance the solution in a direction orthogonal to the mean value of the vector field $\nabla H(y)$. The obvious choice to meet this requirement is to set

$$
(y_1 - y_0) = \frac{h}{2}(J\nabla H(y_0) + J\nabla H(y_1)). \tag{1.63}
$$

We recognize the well known trapezoidal method and the argument above clearly explains why such a method is energy conserving for problem (1.59): the discrete line integral (1.62) induced by the trapezoidal rule matches the original (continuous) line integral (1.61) for linear problems.

This equivalence evaporates if we consider a non-quadratic Hamiltonian function such as

$$
H(y) = H(q, p) = \frac{1}{2}p^2 + q^3. \tag{1.64}
$$

For example, if we use the trapezoidal method to approximate the trajectory of the system defined by (1.64) starting at $y_0 = (1, 0)^\top$, with stepsize $h = 0.1$, the numerical Hamiltonian is no longer constant, as is shown in the second column of Table 1.1. On the other hand, we notice that now the vector field $\nabla H(y)$ evaluated along $\sigma(t)$,

$$
\nabla H(\sigma(t)) = \begin{pmatrix} 3(q_0 + t\frac{q_1 - q_0}{h})^2 \\ p_0 + t\frac{p_1 - p_0}{h} \end{pmatrix},
$$

contains a polynomial of degree two in t as first component. Thus, the Simpson's rule is enough to precisely compute the line integral. By taking into account (1.60) one then obtains:

$$
\begin{aligned}
H(y_1) - H(y_0) &= H(\sigma(h)) - H(\sigma(0)) \\
&= \int_0^h \dot{\sigma}(t)^\top \nabla H(\sigma(t)) dt = (y_1 - y_0)^\top \int_0^1 \nabla H(\sigma(ch)) dc \\
&= (y_1 - y_0)^\top \frac{\nabla H(\sigma(0)) + 4\nabla H(\sigma(h/2)) + \nabla H(\sigma(h))}{6} \\
&= (y_1 - y_0)^\top \frac{\nabla H(y_0) + 4\nabla H\left(\frac{y_0 + y_1}{2}\right) + \nabla H(y_1)}{6}.
\end{aligned}
$$

Arguing as before, we arrive at the method:

$$
y_1 - y_0 = \frac{h}{6}\left[J\nabla H(y_0) + 4J\nabla H\left(\frac{y_0 + y_1}{2}\right) + J\nabla H(y_1) \right]. \tag{1.65}
$$

TABLE 1.1: Errors in the numerical Hamiltonian, when using the s-stage trapezoidal method, with $s = 2, 3$ and stepsize $h = 0.1$, for solving problem (1.64).

t_n	$H(y_n) - H(y_0)$	
	2-stage trapezoidal method	3-stage trapezoidal method
0.0	0	0
0.1	1.6143e-06	0
0.2	4.2746e-05	0
0.3	2.1261e-04	-1.1102e-16
0.4	6.0692e-04	0
0.5	1.2827e-03	-5.5511e-16
0.6	2.2375e-03	-8.8818e-16
0.7	3.4177e-03	-7.7716e-16
0.8	4.7454e-03	-5.5511e-16
0.9	6.1470e-03	-6.6613e-16
1.0	7.5733e-03	-6.6613e-16

This method, called 3-*stage trapezoidal method*, turns out to be energy conserving when the Hamiltonian $H(y)$ is a polynomial of degree at most four.[19] The results presented in the last column of Table 1.1 confirm this property.

It should be now quite clear that the above argument may be easily generalized to tackle the case where $H(y)$ is a polynomial of any arbitrary degree. To compute the line integral without truncation error, we need a quadrature formula with degree of precision at least $\deg H - 1$ (i.e., of order at least $\deg H$). The s-*stage trapezoidal methods* are defined as

$$y_1 - y_0 = h \sum_{i=1}^{s} b_i f\left(y_0 + c_i(y_1 - y_0)\right), \tag{1.66}$$

where $(c_i, b_i)_{i=1,\dots,s}$ are the nodes and weights of the corresponding Newton-Cotes formula, and we have set $f(y) := J\nabla H(y)$ to emphasize that the methods also make sense for general differential equations $\dot y = f(y)$.

The choices $s = 2$ and $s = 3$ lead back to methods (1.63) and (1.65), respectively. As a further example, for $s = 5$, one obtains the 5-stage trapezoidal method,

$$y_1 - y_0 = \frac{h}{90}\left[7f(y_0) + 32f\left(\frac{y_1 + 3y_0}{4}\right) + 12f\left(\frac{y_1 + y_0}{2}\right)\right.$$
$$\left. + 32f\left(\frac{3y_1 + y_0}{4}\right) + 7f(y_1)\right],$$

---

[19]Indeed, the Simpson's rule has degree of precision three.

TABLE 1.2: Energy-conserving properties of methods (1.66) for different choices of the interpolatory quadrature formula.

Abscissae distribution	Newton-Cotes	Lobatto	Gauss
Energy preserving when	$\deg H \leqslant s,\, s+1$	$\deg H \leqslant 2s-2$	$\deg H \leqslant 2s$

which turns out to be energy conserving for polynomial Hamiltonian functions of degree at most 6.

Of course, any other abscissae distribution is also allowed and, actually, recommended to avoid the instability issues of Newton-Cotes formulae for large values of s. Table 1.2 summarizes the behavior of (1.66), for some choices of the abscissae distribution.

Historically, the s-stage trapezoidal methods are the first successful attempt to get an energy preserving integrator based on the idea of the line integral. It has been quite a surprise to discover that the new born formulae actually form a subclass of Runge-Kutta methods with s stages, whose construction is as easy as that of a one-stage formula.[20] Setting

$$\boldsymbol{b} = \begin{pmatrix} b_1 & \cdots & b_s \end{pmatrix}^\top, \qquad \boldsymbol{c} = \begin{pmatrix} c_1 & \cdots & c_s \end{pmatrix}^\top,$$

the vectors containing the weights and abscissae of the given quadrature rule, formulae (1.66) are evidently equivalent to the s-stage Runge-Kutta method with Butcher tableau:

$$\begin{array}{c|c} \boldsymbol{c} & \boldsymbol{c}\boldsymbol{b}^\top \\ \hline & \boldsymbol{b}^\top \end{array}.$$

Routine calculation shows that all s-stage trapezoidal methods are second-order, as is the basic trapezoidal rule, and share the same stability properties of the basic formula.

We notice that the coefficient matrix $\boldsymbol{c}\boldsymbol{b}^\top$ has rank one independently of the choice of s: this reflects the fact that the methods are *mono-implicit*, which means that y_1 is the only unknown to be computed. Indeed, exploiting the property on the rank, the block s-dimensional nonlinear system representing the stages may be recast as a system containing a single unknown. This implies that we can consider any value of s we need, without increasing too much the overall computational cost. We will not go further in the analysis of these methods, since a more general framework will be defined in Chapter 3.

---

[20]This has motivated the name of the methods.

Remark

Equivalence property. Consider a Hamiltonian system,

$$\dot{y} = f(y) := J\nabla H(y),$$

with polynomial Hamiltonian function, and a class of quadrature formulae (e.g. Newton-Cotes, Lobatto, Gauss, etc.). Pick a quadrature formula with s nodes and assume that it is exact if applied to the line integral

$$(y_1 - y_0)^\top \int_0^1 \nabla H(y_0 + c(y_1 - y_0))\mathrm{d}c.$$

It follows that, for all $k \geqslant s$, the corresponding quadrature formulae are also exact and, consequently, the computed approximations y_1, yielded by (1.66) with k in place of s, are all the same up to round-off errors.

In other words, once the integral is computed exactly, *all* the methods are equivalent to each other, and can be formally written as:

$$y_1 - y_0 = h \int_0^1 f(y_0 + c(y_1 - y_0))\mathrm{d}c.$$

1.6 Runge-Kutta line integral methods

The flow of reasoning employed to derive energy-conserving Runge-Kutta methods of order 2 may be easily adapted to devise the class of Runge-Kutta line integral methods of any higher order.

1.6.1 Discrete line integral and collocation methods

As a starting point, we consider the generic collocation method defined at (1.45) applied to a polynomial Hamiltonian system $\dot{y} = J\nabla H(y)$, and the associated solution $v(ch)$ which we now regard as a (polynomial) curve in the phase space $\mathbb{R}^{2m}$

$$c \in [0,1] \mapsto v(ch) \in \mathbb{R}^{2m},$$

joining the initial state vector y_0 to the new approximation y_1 (in fact, $v(0) = y_0$ and $v(h) = y_1$). The line integral of the conservative vector field $\nabla H(y)$ along the curve $v(ch)$ matches the variation of energy at its end points:

$$H(y_1) - H(y_0) = h \int_0^1 \dot{v}(ch)^\top \nabla H(v(ch))\mathrm{d}c. \qquad (1.67)$$

We notice that the integrand $g(c) := \dot{v}(ch)^\top \nabla H(v(ch))$ is a polynomial of degree

$$\deg g \;=\; s - 1 + (\deg H - 1)s \;=\; s \cdot \deg H - 1. \tag{1.68}$$

Assume that the underlying quadrature formula $(c_i, b_i)_{i=1,\dots,s}$ defining the collocation method has order p (thus its degree of precision is $p-1$). Evaluating the integral by this quadrature rule yields

$$H(y_1) - H(y_0) \;=\; h \sum_{i=1}^{s} b_i \dot{v}(c_i h)^\top \nabla H(v(c_i h)) + E_s, \tag{1.69}$$

where the first term at right-hand side is the discrete line integral induced by the method while the last term is the error introduced by the quadrature formula: $E_s = C h^{p+1} g^{(p)}(\xi)$, with C a constant independent of h and $\xi \in [0,1]$. It is easy to check that the discrete line integral appearing in (1.69) vanishes because of the collocation conditions (1.45):

$$\dot{v}(c_i h)^\top \nabla H(v(c_i h)) \;=\; \nabla H(v(c_i h))^\top J^\top \nabla H(v(c_i h)) \;=\; 0.$$

This result reveals an interesting analogy between the behavior of the continuous line integral evaluated along the true solution $y(t)$ and the discrete line integral evaluated along its numerical approximation: both quantities vanish. If the quadrature were exact, we could deduce the energy-conservation property of the numerical method for the problem at hand. From (1.68) we see that

$$E_s = 0 \quad \Longleftrightarrow \quad g^{(p)}(c) \equiv 0 \quad \Longleftrightarrow \quad p \geqslant s \cdot \deg H.$$

Now, the maximum achievable order for an interpolatory quadrature formula is $p = 2s$ and corresponds to the Gauss-Legendre quadrature rule. For such a method we get $E_s = 0$ under the assumption that $\deg H \leqslant 2$: we have obtained the well-known result that Gauss collocation methods conserve quadratic Hamiltonians while failing to conserve polynomial Hamiltonian functions of higher degree.[21]

1.6.2   Derivation of Runge-Kutta LIMs

This premise gives us two enlightening hints about how to define formulae capable of retrieving the conservation property $H(y_1) = H(y_0)$ for any high degree polynomial $H(y)$:

---

[21] That Lobatto IIIA methods are also energy-preserving for linear problems cannot be deduced from (1.69). However, it is possible to prove that such methods may be as well defined by means of a quasi-collocation polynomial $\sigma(ch)$ whose degree is $s - 1$ instead of s and a similar computation as that presented here may be employed to deduce the energy-conservation property of these formulae for linear systems (see [37, 48] for details).

1. First of all, the flow of reasoning considered above may be reversed. We could start by considering an unknown polynomial curve of degree s, $v(ch)$, in the phase space $\mathbb{R}^{2m}$, and the line integral $\int_v \nabla H(y) \cdot y$ defined at (1.67). Then, we should try to approximate the integral by means of an exact quadrature formula. Choosing an interpolatory quadrature rule based at s nodes $c_1, \ldots, c_s \in [0, 1]$, because of (1.68), we conclude that the only possible choice to get a discrete line integral matching the continuous one is to handle a linear problem ($\deg H = 2$) and to consider a Gauss-Legendre distribution of the nodes c_i. In fact, excluding the trivial case of linear Hamiltonian functions, this is doomed to be the only combination yielding

$$H(y_1) - H(y_0) \;=\; h \sum_{i=1}^{s} b_i \dot{v}(c_i h)^\top \nabla H(v(c_i h)),$$

where, of course, $y_0 := v(0)$ and $y_1 := v(h)$. Thus, to get energy conservation, it is sufficient to impose the orthogonality conditions

$$\dot{v}(c_i h) = J \nabla H(v(c_i h)), \quad i = 1, \ldots, s, \tag{1.70}$$

along with

$$v(0) = y_0, \tag{1.71}$$

and we would be led back to the Gauss-Legendre collocation method (1.45).

2. We observe that the line integral in (1.67) may be thought of as the solution at time $t_1 = t_0 + h$ of the pure quadrature initial value problem

$$\dot{z} \;=\; g(t) := \dot{v}(t)^\top \nabla H(v(t)), \qquad z(0) = 0, \tag{1.72}$$

while the related discrete line integral in (1.69) is nothing but the solution at time $t_1 = h$ computed by the collocation method (1.70)-(1.71).

The first remark suggests that using a quadrature interpolatory rule defined on $s = \deg v$ abscissae is a stumbling block that limits the Hamiltonian function to be at most quadratic. The second remark says that, when we evaluate the line integral by means of its discrete counterpart, we are actually performing one step of the collocation method applied to problem (1.72).

The trick is then to define a method that has order p when applied to a general differential equation $\dot{y} = f(t, y)$ and order $d \geqslant p$ when applied to a pure quadrature problem $\dot{z} = g(t)$ as is the problem of approximating the line integral.

With this purpose in mind, we now try to perform the computation outlined at item 1 again, but now choosing a polynomial $\sigma(ch)$, of degree s, such that $\sigma(0) = y_0$ and a quadrature rule $(c_i, b_i)_{i=1,\ldots,k}$ based at k (rather than s) nodes. It should be argued that the degree of $\sigma(ch)$ will influence the order

p of the method while choosing k high enough will provide us with an exact quadrature formula when solving the polynomial problem (1.72).

Let us consider a polynomial Hamiltonian system $\dot{y} = J\nabla H(y)$ and set $\nu = \deg H(y)$. Let $\{P_j(c)\}_{j=0,\ldots,s-1}$ be a basis for the s-dimensional space Π_{s-1} of polynomials with degree at most $s-1$ defined on $[0,1]$. The polynomial $\dot{\sigma}(ch)$ expanded on the given basis reads

$$\dot{\sigma}(ch) = \sum_{j=0}^{s-1} P_j(c)\gamma_j, \qquad (1.73)$$

where $\gamma_j \in \mathbb{R}^{2m}$ are unknown vector coefficients. Integrating both sides of (1.73) with respect to the variable c gives

$$\sigma(ch) = y_0 + h\sum_{j=0}^{s-1} \int_0^c P_j(\tau)\mathrm{d}\tau\,\gamma_j, \qquad (1.74)$$

and, as usual, we set $y_1 := \sigma(h)$. We now assume that the order d of the quadrature formula $(c_i, b_i)_{i=1,\ldots,k}$ satisfies the condition $d \geqslant \nu s$: for example, if we choose a Gauss-Legendre distribution of the nodes, then $d = 2k$ and k should be such that $k \geqslant \nu s/2$. By virtue of (1.68), this means that the discrete line integral is exact, so we may write

$$
\begin{aligned}
H(y_1) - H(y_0) &= h\int_0^1 \dot{\sigma}(ch)^\top \nabla H(\sigma(ch))\mathrm{d}c \\
&= h\sum_{i=1}^{k} b_i \dot{\sigma}(c_i h)^\top \nabla H(\sigma(c_i h)) \\
&= h\sum_{i=1}^{k} b_i \sum_{j=0}^{s-1} P_j(c_i)\gamma_j^\top \nabla H(\sigma(c_i h)) \\
&= h\sum_{j=0}^{s-1} \gamma_j^\top \sum_{i=1}^{k} b_i P_j(c_i)\nabla H(\sigma(c_i h)).
\end{aligned}
$$

Energy conservation is gained by imposing $H(y_1) - H(y_0) = 0$, i.e., that the free coefficients γ_j satisfy, for $j = 0,\ldots,s-1$, the orthogonality conditions

$$
\begin{aligned}
\gamma_j &= \eta_j J \sum_{i=1}^{k} b_i P_j(c_i)\nabla H(\sigma(c_i h)) \\
&= \eta_j \sum_{i=1}^{k} b_i P_j(c_i) J\nabla H(\sigma(c_i h)) \\
&= \eta_j \sum_{i=1}^{k} b_i P_j(c_i) J\nabla H\left(y_0 + h\sum_{r=0}^{s-1} \int_0^{c_i} P_r(\tau)\mathrm{d}\tau\,\gamma_r\right), \qquad (1.75)
\end{aligned}
$$

where η_j are free real coefficients that should be determined by imposing the consistency conditions (see Section 1.6.6). It is worth mentioning that the nonlinear system (1.75) has block dimension s whichever is k: after solving (1.75) we may use (1.74) to advance the solution by setting

$$y_1 \;=\; y_0 + h \sum_{j=0}^{s-1} \int_0^1 P_j(\tau)\mathrm{d}\tau\, \gamma_j, \tag{1.76}$$

where $\int_0^1 P_j(\tau)\mathrm{d}\tau$ are known constants once the basis has been assigned.

Remark

Under the current assumption that the quadrature rule $(c_i, b_i)_{i=1,\dots,k}$ be exact, we can formally retain the integrals in place of the sums during the computation to derive the coefficients γ_j. In so doing, we would get

$$H(y_1) - H(y_0) \;=\; h \sum_{j=0}^{s-1} \gamma_j^\top \int_0^1 P_j(c)\nabla H(\sigma(ch))\mathrm{d}c,$$

and the orthogonality conditions (1.75) would become

$$\gamma_j \;=\; \eta_j \int_0^1 P_j(c) J \nabla H \left(y_0 + h \sum_{r=0}^{s-1} \int_0^c P_r(\tau)\mathrm{d}\tau\, \gamma_r \right) \mathrm{d}c \tag{1.77}$$

We emphasize that:

1. (1.75) and (1.77) identify the very same system if the underlying quadrature rule is exact;

2. from a theoretical point of view, (1.77) may be viewed as the limit formulae, as $k \to \infty$, if $H(y)$ is not a polynomial in y.

1.6.3   What have we got?

Actually, what we have got from (1.75)-(1.76) is a Runge-Kutta method in disguise. To see this, we define the vectors

$$Y_i := \sigma(c_i h), \qquad i = 1, \dots, k,$$

which, as will be clear in a while, will be the stages of an equivalent Runge-Kutta formula. From (1.74) and (1.75) we obtain

$$
\begin{aligned}
Y_i &= y_0 + h \sum_{r=0}^{s-1} \int_0^{c_i} P_r(\tau)\mathrm{d}\tau\, \gamma_r \\
&= y_0 + h \sum_{r=0}^{s-1} \int_0^{c_i} P_r(\tau)\mathrm{d}\tau\, \eta_r \sum_{j=1}^{k} b_j P_r(c_j) J\nabla H(Y_j) \\
&= y_0 + h \sum_{j=1}^{k} b_j \left(\sum_{r=0}^{s-1} \eta_r P_r(c_j) \int_0^{c_i} P_r(\tau)\mathrm{d}\tau \right) J\nabla H(Y_j), \quad i = 1,\ldots,k,
\end{aligned}
$$

while (1.76) becomes

$$
y_1 = y_0 + h \sum_{j=1}^{k} b_j \left(\sum_{r=0}^{s-1} \eta_r P_r(c_j) \int_0^1 P_r(\tau)\mathrm{d}\tau \right) J\nabla H(Y_j).
$$

In conclusion, the Runge-Kutta line integral method we have derived is defined by the following tableau:

$$
\begin{array}{c|c}
\begin{matrix} c_1 \\ \vdots \\ c_k \end{matrix} & A := (\alpha_{ij}) \\
\hline
 & \beta_1 \cdots \beta_k
\end{array}, \tag{1.78}
$$

$$
\begin{aligned}
\alpha_{ij} &:= b_j \sum_{r=0}^{s-1} \eta_r P_r(c_j) \int_0^{c_i} P_r(\tau)\mathrm{d}\tau, \\
\beta_j &:= b_j \sum_{r=0}^{s-1} \eta_r P_r(c_j) \int_0^1 P_r(\tau)\mathrm{d}\tau, \quad i,j = 1,\ldots,k.
\end{aligned}
$$

An example of such a method is reported at the end of Section 1.6.5.

1.6.4 Fundamental and silent stages

That the k-dimensional block system (1.78) is equivalent to the s-dimensional block system (1.75) implies that matrix A is singular with rank equal to s, independently of the value of k. A formal proof exploiting the algebraic properties of matrix A will be supplied in Theorem 3.8. For a more informal explanation, we notice that, by definition, Y_i are the values attained by the polynomial $\sigma(ch)$ at the abscissae c_i. Since the degree of σ is s, it will be completely identified by s (rather than k) stages plus the point y_0. For example, we could express $\sigma(ch)$ in terms of the first s stages as

$$
\sigma(ch) = \ell_0(c)y_0 + \sum_{j=1}^{s} \ell_j(c)Y_j, \tag{1.79}
$$

where $\ell_j(c)$ denotes the j-th Lagrange polynomial defined on the nodes $0, c_1, \ldots, c_s$.[22] From (1.79) we deduce that the remaining $k - s$ stages may be expressed as a linear combination of the first s stages:

$$Y_i = \ell_0(c_i)y_0 + \sum_{j=1}^{s} \ell_j(c_i)Y_j, \qquad i = s+1, \ldots, k. \qquad (1.80)$$

After setting

$$Z := \begin{pmatrix} Y_1 \\ \vdots \\ Y_s \end{pmatrix}, \qquad W := \begin{pmatrix} Y_{s+1} \\ \vdots \\ Y_k \end{pmatrix}, \qquad (1.81)$$

and

$$a_0 := \begin{pmatrix} \ell_0(c_{s+1}) \\ \vdots \\ \ell_0(c_k) \end{pmatrix}, \qquad M := \begin{pmatrix} \ell_1(c_{s+1}) & \cdots & \ell_s(c_{s+1}) \\ \vdots & & \vdots \\ \ell_1(c_k) & \cdots & \ell_s(c_k) \end{pmatrix},$$

we have, in matrix notation,

$$W = a_0 \otimes y_0 + M \otimes I \, Z. \qquad (1.82)$$

In conclusion, we can cast the nonlinear system defining the stages as

$$\begin{pmatrix} I_s & O \\ -M & I_{k-s} \end{pmatrix} \otimes I \begin{pmatrix} Z \\ W \end{pmatrix}$$

$$= \begin{pmatrix} \mathbf{1}_s \\ a_0 \end{pmatrix} \otimes y_0 + h \begin{pmatrix} A_{11} & A_{12} \\ O & O \end{pmatrix} \otimes I \begin{pmatrix} f(Z) \\ f(W) \end{pmatrix}, \qquad (1.83)$$

where the block matrix $\begin{pmatrix} A_{11} & A_{12} \end{pmatrix}$ contains the first s rows of the coefficient matrix A defined at (1.78). Compared with the system

$$Y = \mathbf{1}_k \otimes y_0 + hA \otimes I \, f(Y), \qquad (1.84)$$

associated with the tableau (1.78), we see that now the linear and nonlinear components of the system are completely uncoupled, so (1.83) turns out to be more advantageous during the implementation of the methods.[23]

To outline that only s stages contribute the nonlinear part of system (1.84), with reference to (1.81)–(1.83), we call $Y_1, \ldots, Y_s$ *fundamental stages*, and $Y_{s+1}, \ldots, Y_k$, *silent stages*. Of course, the fundamental stages need not necessarily be the first s ones and in fact, to improve the conditioning of the system, it is more convenient to let the fundamental and silent stages be interlaced.

---

[22] For simplicity, we are now assuming that $c_1 > 0$.

[23] The implementation details of Runge-Kutta line integral methods are discussed in Chapter 4, together with a number of numerical illustrations.

1.6.5 Hamiltonian Boundary Value Methods

Besides the formulations (1.75), (1.78), and (1.83), there is a further, different shape that line integral methods of Runge-Kutta type may assume. Historically, the formulation we are going to introduce is tied to the first successful attempts to devise energy preserving methods based on the discrete line integral approach, so it has influenced the name conferred to this class of methods: HBVMs, which stands for *Hamiltonian Boundary Value Methods*. Here we want to explore this slightly different path of investigation.

Consider again the relation between the stages Y_i and the vector γ_j:

$$Y_i = y_0 + h \sum_{r=0}^{s-1} \int_0^{c_i} P_r(\tau) \mathrm{d}\tau \, \gamma_r, \qquad i = 1, \ldots, k.$$

If we collect the first s such equations, we obtain the nonlinear system

$$Z = 1_s \otimes y_0 + h \mathcal{I}_s \otimes I \, \gamma,$$

where (see (1.81)), Z is the block vector containing the stages $Y_1, \ldots, Y_s$,

$$\gamma = \begin{pmatrix} \gamma_0 \\ \vdots \\ \gamma_{s-1} \end{pmatrix},$$

while matrix

$$\mathcal{I}_s = \left(\int_0^{c_i} P_{r-1}(\tau) \mathrm{d}\tau \right)_{i, r = 1, \ldots, s}$$

is formally similar to that introduced in (1.55), but now the basis $\{P_r\}$ is generic. Assuming that matrix $\mathcal{I}_s$ is nonsingular, one may recast the coefficient vectors γ_j as linear combinations of the first s stages Y_i. Consequently, the orthogonal conditions (1.75) may be imposed directly on such linear combinations rather than on the vectors γ_j, thus obtaining a nonlinear system having the stages as unknowns, which is evidently equivalent to (1.78) and (1.83).

If introducing the strategy of recasting the γ_j in terms of the stages before imposing the orthogonality conditions may sound a bit stretched, it should be noticed that, in some situations, these linear combinations are already available at the beginning of the construction procedure, and may be used right away. This is the case, for example, if we start the computation by considering the expansion of the polynomial $\sigma(ch)$ rather than of its derivative. In fact, the interpolation conditions $\sigma(c_i h) = Y_i$, $i = 1, \ldots, s$, could be immediately imposed, which would cause the stages Y_i to enter the expression of $\dot{\sigma}(c_i h)$ before imposing the orthogonality conditions (1.75).

To elucidate this aspect, we consider an explicit example which introduces one of the first line integral formulae of order higher than two in the literature, derived by expanding $\sigma(ch)$ along the Newton polynomial basis. Consider the Lobatto quadrature rule of order $d = 8$ defined by the following five nodes $c_i \in [0,1]$ and corresponding weights b_i:

$$(c_1, c_2, c_3, c_4, c_5) \quad := \quad \left(0, \frac{1}{2} - \frac{1}{14}\sqrt{21}, \frac{1}{2}, \frac{1}{2} + \frac{1}{14}\sqrt{21}, 1\right), \quad (1.85)$$

$$(b_1, b_2, b_3, b_4, b_5) \quad := \quad \left(\frac{1}{20}, \frac{49}{180}, \frac{16}{45}, \frac{49}{180}, \frac{1}{20}\right). \quad (1.86)$$

Our aim is to construct a line integral method of order $p = 4$, so we let $\sigma(ch)$ be a polynomial of degree $s = 2$. Since the energy-conservation property requires $d \geqslant s \deg H$, the resulting formula will be appropriate for polynomial Hamiltonian functions with degree up to four, so we assume $\deg(H(p,q)) \leqslant 4$. We choose $\sigma(ch)$ as the Newton polynomial that interpolates the data $(c_1, Y_1), (c_3, Y_3)$ and (c_5, Y_5).[24] This means that, in the final formula, Y_1, Y_3, Y_5, will act as the fundamental stages, while Y_2 and Y_4 will be the silent stages. Using appropriate divided differences we obtain

$$\sigma(ch) \quad := \quad Y_1 + (Y_5 - Y_1)c + 2(Y_5 - 2Y_3 + Y_1)c(c-1), \quad (1.87)$$

and hence

$$h\,\dot{\sigma}(ch) \quad = \quad (Y_5 - Y_1) + 2(Y_5 - 2Y_3 + Y_1)(2c-1).$$

According to (1.80), the two additional stages Y_2 and Y_4 take the form

$$Y_2 \quad := \quad \sigma(c_2 h) = \frac{1}{14}(3 + \sqrt{21})Y_1 + \frac{4}{7}Y_3 + \frac{1}{14}(3 - \sqrt{21})Y_5, \quad (1.88)$$

and

$$Y_4 \quad := \quad \sigma(c_4 h) = \frac{1}{14}(3 - \sqrt{21})Y_1 + \frac{4}{7}Y_3 + \frac{1}{14}(3 + \sqrt{21})Y_5. \quad (1.89)$$

The variation of energy at the end points of σ is:

$$
\begin{aligned}
H(y_1) - H(y_0) \quad &= \quad h \int_0^1 \dot{\sigma}(ch)^\top \nabla H(\sigma(ch))\,dc \\
&= \quad (Y_5 - Y_1)^\top \int_0^1 \nabla H(\sigma(ch))\,dc \\
&\quad + 2(Y_5 - 2Y_3 + Y_1)^\top \int_0^1 (2c-1)\nabla H(\sigma(ch))\,dc.
\end{aligned}
$$

Since $\deg H \leqslant 4$ and $\deg \sigma = 2$, the two integrand functions have degree

---

[24] Since $c_1 = 0$ and $c_5 = 1$, it follows that $Y_1 = y_0$ and $Y_5 = y_1$.

less than or equal to 7 and therefore, as we had anticipated, our Lobatto quadrature formula will compute them exactly. Consequently, we have:

$$H(y_1) - H(y_0) = (Y_5 - Y_1)^\top \sum_{i=1}^{5} b_i \nabla H(Y_i)$$

$$+ 2(Y_5 - 2Y_3 + Y_1)^\top \sum_{i=1}^{5} b_i(2c_i - 1)\nabla H(Y_i).$$

Requiring that $H(y_1) = H(y_0)$ results in the following two orthogonality conditions involving the stages, which correspond to (1.75):

$$Y_5 - Y_1 = h\eta_0 \sum_{i=1}^{5} b_i J \nabla H(Y_i) \tag{1.90}$$

and

$$Y_5 - 2Y_3 + Y_1 = h\eta_1 \sum_{i=1}^{5} b_i(2c_i - 1) J \nabla H(Y_i). \tag{1.91}$$

The above conditions tell us that the stages must satisfy two linear multistep formulae: their order is maximized by choosing $\eta_0 = 1$ and $\eta_1 = \frac{3}{2}$.[25] The resulting conservative method is the collection of the linear multistep formulae (1.88)–(1.91). It is defined by the following nonlinear system,

$$\left(\, a_0 \, | \, A \, \right) \otimes I \, Y = h \left(\, b_0 \, | \, B \, \right) \otimes I \, f(Y), \tag{1.92}$$

where [26]

$$Y^\top := \left(\, y_0^\top \quad Y_2^\top \quad Y_3^\top \quad Y_4^\top \quad y_1^\top \, \right), \qquad f(Y) := J \nabla H(Y),$$

and the two coefficient matrices have entries:

$$\left(\, a_0 \, | \, A \, \right) := \begin{pmatrix} -\frac{1}{14}(\sqrt{21} + 3) & 1 & -\frac{4}{7} & 0 & \frac{1}{14}(\sqrt{21} - 3) \\ 1 & 0 & -2 & 0 & 1 \\ \frac{1}{14}(\sqrt{21} - 3) & 0 & -\frac{4}{7} & 1 & -\frac{1}{14}(\sqrt{21} + 3) \\ -1 & 0 & 0 & 0 & 1 \end{pmatrix}, \tag{1.93}$$

and

$$\left(\, b_0 \, | \, B \, \right) := \begin{pmatrix} 0 & 0 & 0 & 0 & 0 \\ -\frac{3}{40} & -\frac{7}{120}\sqrt{21} & 0 & \frac{7}{120}\sqrt{21} & \frac{3}{40} \\ 0 & 0 & 0 & 0 & 0 \\ \frac{1}{20} & \frac{49}{180} & \frac{16}{45} & \frac{49}{180} & \frac{1}{20} \end{pmatrix}. \tag{1.94}$$

---

[25] See Section 1.6.6 for details.
[26] Notice the presence of the initial and final state vectors y_0 and y_1 in Y, in place of Y_1 and Y_5.

One-step methods such as (1.92), obtained as a collection of linear multi-step formulae, may be contextualized within the framework of block one-step methods or *block Boundary Value Methods (block-BVMs)*. The name of this latter class of methods stems from the fact that the discrete problem they define can be viewed, locally in the time interval $[0, h]$, as a natural integrator for Boundary Value Problems. In fact, as is the case with boundary value problems, the solution of the associated nonlinear system (1.92) may be uniquely defined by providing a boundary condition involving y_0 and y_1 in the form $g(y_0, y_1) = 0$, for a suitable function g. Of course, providing y_0 results in an integrator for initial value problems, as is the use of formula (1.92) in the present context. For such a reason, line integral methods such as (1.92) were consistently called Hamiltonian Boundary Value Methods (HBVMs) and such a name propagated to all the equivalent formulation we have discussed previously.

It should be self-evident that a general s-stage block-BVM may be always recast in Runge-Kutta form. First of all, we see that the row-sums of the matrix $\begin{pmatrix} a_0 & | & A \end{pmatrix} \in \mathbb{R}^{s-1 \times s}$ are all null. This follows from the pre-consistency condition: in the present context we can justify this property by observing that each divided difference with at least two arguments computed along the constant values $Y_i = 1$ is null, and consequently $\sigma(ch) \equiv Y_1$. Hence we have $-A^{-1}a_0 = 1_{s-1}$. Now, pulling the column a_0 out of the left hand-side coefficient matrix in (1.92) yields

$$A \otimes I \hat{Y} \;\; = \;\; -a_0 \otimes y_0 + h \begin{pmatrix} b_0 & | & B \end{pmatrix} \otimes I f(Y), \qquad (1.95)$$

where

$$Y^\top := \begin{pmatrix} Y_1^\top & | & \hat{Y}^\top \end{pmatrix} := \begin{pmatrix} Y_1^\top & | & Y_2^\top & \cdots & Y_s^\top \end{pmatrix}.$$

Multiplying both sides of (1.95) by A^{-1} allows us to recast (1.92) in the form

$$Y \;\; = \;\; 1_s \otimes y_0 + h \begin{pmatrix} 0 \dots\dots 0 \\ A^{-1}(b_0|B) \end{pmatrix} \otimes I f(Y). \qquad (1.96)$$

Equation (1.96) is a s-stage Runge-Kutta method with the coefficients b_i, $i = 1, \dots, s$, taken from the last row of the matrix $A^{-1}(b_0|B)$. Following this computation, the HBVM (1.92)–(1.94) is recast as the Runge-Kutta method defined by means of the Butcher tableau in Table 1.3. The order four can be stated by checking, for example, the well-known simplifying conditions.[27]

Conversely, a Runge-Kutta line integral method applied to a polynomial Hamiltonian system may always be changed to an equivalent block-BVM: in general, it will suffice to add a zero leading row, if the abscissa 0 is not present, and/or include the quadrature as the last row of the Butcher matrix, if the abscissa 1 is not present.

---

[27]See, for example, [107, Theorem 5.1, page 71].

TABLE 1.3: Butcher tableau of the HBVM (1.92)–(1.94).

$$
\begin{array}{c|ccccc}
0 & 0 & 0 & 0 & 0 & 0 \\[4pt]
\frac{1}{2} - \frac{1}{14}\sqrt{21} & \frac{13}{280} - \frac{1}{280}\sqrt{21} & \frac{49}{360} - \frac{1}{360}\sqrt{21} & \frac{8}{45} - \frac{8}{315}\sqrt{21} & \frac{49}{360} - \frac{13}{360}\sqrt{21} & \frac{1}{280} - \frac{1}{280}\sqrt{21} \\[4pt]
\frac{1}{2} & \frac{1}{16} & \frac{49}{360} + \frac{7}{240}\sqrt{21} & \frac{8}{45} & \frac{49}{360} - \frac{7}{240}\sqrt{21} & -\frac{1}{80} \\[4pt]
\frac{1}{2} + \frac{1}{14}\sqrt{21} & \frac{13}{280} + \frac{1}{280}\sqrt{21} & \frac{49}{360} + \frac{13}{360}\sqrt{21} & \frac{8}{45} + \frac{8}{315}\sqrt{21} & \frac{49}{360} + \frac{1}{360}\sqrt{21} & \frac{1}{280} + \frac{1}{280}\sqrt{21} \\[4pt]
1 & \frac{1}{20} & \frac{49}{180} & \frac{16}{45} & \frac{49}{180} & \frac{1}{20} \\[4pt]
\hline
& \frac{1}{20} & \frac{49}{180} & \frac{16}{45} & \frac{49}{180} & \frac{1}{20}
\end{array}
$$

Remark

In consideration of the above analysis, in the sequel of the manuscript we will use indifferently the names *Runge-Kutta Line Integral Method* or *Hamiltonian Boundary Value Method* to denote the very same method, independently of the formulation adopted:

Runge-Kutta Line Integral Methods

↓↑

Hamiltonian Boundary Value Methods

1.6.6   Choice of the basis

We consider again (1.73) and (1.75):

$$\dot{\sigma}(ch) = \sum_{j=0}^{s-1} P_j(c)\gamma_j, \tag{1.97}$$

$$\gamma_j = \eta_j \sum_{i=1}^{k} b_i P_j(c_i) f\left(c_i h,\ y_0 + h \sum_{r=0}^{s-1} \int_0^{c_i} P_r(\tau)\mathrm{d}\tau\ \gamma_r\right), \tag{1.98}$$

where now we let the method be applied to a generic initial value problem $\dot{y} = f(t, y)$. As we had anticipated in Section 1.6.2, the free scalar coefficients η_j are generally computed by imposing the consistency condition that the method is exact if applied to the problem $\dot{y} = 1$. This means that we must require $\dot{\sigma}(ch) \equiv 1$.[28] Since $\deg P_j(c) \leqslant s-1$, the quadrature rule $(c_i, b_i)_{i=1,\dots,k}$, with $k \geqslant s$, turns out to be exact for the problem at hand, so we obtain the following explicit expression for γ_j:

$$\gamma_j \;=\; \eta_j \sum_{i=1}^{k} b_i P_j(c_i) \;=\; \eta_j \int_0^1 P_j(c)\mathrm{d}c, \qquad j = 0, \dots, s-1.$$

Thus, from (1.97) we deduce that the consistency condition yields the following equation in the unknowns η_j:

$$\sum_{j=0}^{s-1} \int_0^1 P_j(x)\mathrm{d}x\, P_j(c)\eta_j \;=\; 1. \tag{1.99}$$

To concretely compute the η_j, we could fix a set of s distinct abscissae $\hat{c}_i \in [0, 1]$ and impose (1.99) for each choice $c = \hat{c}_i$, $i = 1, \dots, s$. In so doing, we

---

[28] From (1.97)-(1.98) and the condition $\sigma(0) = y_0$, we see that for $f = 0$ the method gives the exact solution.

obtain the system

$$\mathcal{P}_s \mathcal{D}_s \boldsymbol{\eta} \;=\; \mathbf{1},\tag{1.100}$$

with

$$\mathcal{P}_s \;=\; \begin{pmatrix} P_0(\hat{c}_1) & \cdots & P_{s-1}(\hat{c}_1) \\ \vdots & & \vdots \\ P_0(\hat{c}_s) & \cdots & P_{s-1}(\hat{c}_s) \end{pmatrix} \in \mathbb{R}^{s\times s}, \qquad \mathbf{1} = \begin{pmatrix} 1 \\ \vdots \\ 1 \end{pmatrix} \in \mathbb{R}^s,$$

$$\mathcal{D}_s \;=\; \begin{pmatrix} \int_0^1 P_0(x)\mathrm{d}x & & \\ & \ddots & \\ & & \int_0^1 P_{s-1}(x)\mathrm{d}x \end{pmatrix} \in \mathbb{R}^{s\times s}, \qquad \boldsymbol{\eta} = \begin{pmatrix} \eta_0 \\ \vdots \\ \eta_{s-1} \end{pmatrix} \in \mathbb{R}^s.$$

Since $\{P_j(c)\}_{j=0,\ldots,s-1}$ form a basis for Π_{s-1}, we see that matrix $\mathcal{P}_s$ is always invertible, because the abscissae $\hat{c}_i$ are distinct, thus the unique solvability of (1.100) depends on the invertibility of the diagonal matrix $\mathcal{D}_s$ and, hence, on the specific basis introduced. We consider a few examples.

Power basis

With the choice

$$P_j(c) := c^j, \qquad j = 0, \ldots, s-1,$$

matrix $\mathcal{P}_s$ becomes the well-known Vandermonde matrix defined on the nodes $\hat{c}_i$:

$$\mathcal{P}_s \;=\; \begin{pmatrix} 1 & \hat{c}_1 & \cdots & \hat{c}_1^{s-1} \\ \vdots & \vdots & & \vdots \\ 1 & \hat{c}_{s-1} & \cdots & \hat{c}_s^{s-1} \end{pmatrix},$$

so $\mathcal{P}_s^{-1}\mathbf{1} = \boldsymbol{e}_1$, with $\boldsymbol{e}_1 := (1,0,\ldots,0)^\top$. The solution of system (1.100) is then

$$\eta_0 \;=\; 1, \qquad \eta_j \;=\; 0, \quad \text{for} \quad j = 1, \ldots, s-1.$$

It follows that $\sigma(ch)$ must be a polynomial of degree one, so this basis is not suitable for deriving higher order methods.

Lagrange basis

Let us fix a set of s distinct nodes $\hat{c}_i \in [0,1]$ such that the weights $\hat{b}_i$ of the corresponding interpolatory quadrature formula are nonvanishing: $\hat{b}_i \neq 0$, $i = 1, \ldots, s$. We consider the choice

$$P_{j-1}(c) := \ell_j(c) \equiv \prod_{i\neq j} \frac{c - \hat{c}_i}{\hat{c}_j - \hat{c}_i}, \qquad j = 1, \ldots, s.$$

Since $\ell_j(\hat{c}_i) = \delta_{ij}$ (the Kronecker symbol), matrix $\mathcal{P}_s$ becomes the identity matrix, while matrix $\mathcal{D}_s$ has $\hat{b}_1, \hat{b}_2, \ldots, \hat{b}_s$ as diagonal entries. It follows that (1.100) has a unique solution given by

$$\eta_{i-1} = \frac{1}{\hat{b}_i}, \qquad i = 1, \ldots, s.$$

Newton basis

An example of use of this basis has been discussed in Section 1.6.5. Here we consider it again to show how to tackle the problem of finding the coefficients η_i in the event that the consistency condition leading to equation (1.99) is not enough to determine all of them. We write again the two orthogonality conditions (1.90) and (1.91),

$$Y_5 - Y_1 = h\eta_0 \sum_{i=1}^{5} b_i f(c_i h, Y_i), \qquad (1.101)$$

$$Y_5 - 2Y_3 + Y_1 = h\eta_1 \sum_{i=1}^{5} b_i (2c_i - 1) f(c_i h, Y_i), \qquad (1.102)$$

where $Y_i = \sigma(c_i h)$, $i = 1, \ldots, 5$. Setting $f = 1$, and considering the values of the abscissae c_i and weights b_i listed at (1.85) and (1.86), we get

$$Y_5 - Y_1 = h\eta_0 \sum_{i=1}^{5} b_i = h\eta_0,$$

$$Y_5 - 2Y_3 + Y_1 = h\eta_1 \sum_{i=1}^{5} b_i (2c_i - 1) = 0.$$

Imposing that the computed solution is exact, namely $\sigma(ch) = y_0 + ch$, requires that $Y_i = y_0 + c_i h$ which, in turn, implies $\eta_0 = 1$ as is easily checked.

The coefficient η_1 remains free and must be determined by imposing a different condition. The natural choice is to try to increase the order by requiring that the method be exact if applied to

$$\dot{y} = c, \qquad c \in [0,1], \qquad \text{with} \qquad y(0) = 0,$$

whose solution is $y(c) = \frac{1}{2}c^2$. Choosing a stepsize $h = 1$ and substituting $f(ch, y) = f(c, y) = c$ in (1.101) and (1.102) gives

$$Y_5 - Y_1 = \sum_{i=1}^{5} b_i c_i = \int_0^1 c\,dc = \frac{1}{2},$$

$$Y_5 - 2Y_3 + Y_1 = \eta_1 \sum_{i=1}^{5} b_i c_i (2c_i - 1) = \eta_1 \int_0^1 c(2c - 1)dc = \frac{1}{6}\eta_1.$$

Since $Y_1 = y_0 = y(0) = 0$, the first equation gives the correct result $Y_5 = \frac{1}{2} \equiv y(1)$. We finally introduce this value in the second equation, and impose the condition $Y_3 = y(\frac{1}{2}) = \frac{1}{8}$, thus obtaining $\eta_1 = \frac{3}{2}$, as was stated during the construction of the method in Section 1.6.5.

Legendre basis

The shape of Runge-Kutta line integral formulae defined with the aid of the Lagrange basis and the Newton basis will depend on the specific distribution of the nodes $\hat{c}_i$, $i = 1, \ldots, s$. A change of the location of these nodes will result in a change of the coefficients defining the method. Excluding the energy conservation, which is an inherent feature of the construction procedure, the choice of the nodes $\hat{c}_i$ will influence a number of properties of the corresponding method, among which is its order of convergence. For example, it is not just by chance that the polynomial $\sigma(ch)$ defined at (1.87) has been constructed in such a way to be symmetric on the interval $[0, 1]$: the resulting line integral method (1.93)-(1.94), defined along the Newton basis, would have not reached order four if we had chosen a different set of fundamental stages to define $\sigma(ch)$.

The above remarks suggest the use of a polynomial basis $\{P_j(c)\}$ made up of polynomials which are symmetric on the interval $[0, 1]$ and whose definition does not require the use of the nodes $\hat{c}_i$ (as in the case, e.g., of the Lagrange basis). We have already encountered a family of polynomials sharing these features in Section 1.4.3: we are talking of the shifted and normalized Legendre polynomials

$$P_j(c) = \frac{\sqrt{2j+1}}{j!} \frac{\mathrm{d}^j}{\mathrm{d}c^j} \left[(c^2 - c)^j \right], \qquad j = 0, \ldots, s-1,$$

which form an orthonormal basis for Π_{s-1}, the linear space of polynomials of degree at most $s - 1$.[29]

Since $P_0(c) = 1$ and, because of the orthogonality conditions, $\int_0^1 P_j(x)\mathrm{d}x = 0$, $j = 1, \ldots, s-1$, equation (1.99) yields $\eta_0 = 1$, while the other coefficients η_j remain free. We can determine them by imposing that the method be exact if applied to a quadrature problem of the form $\dot{y} = p(c)$ where $p(c)$ is a generic polynomial of degree not exceeding $s - 1$. This is tantamount to require that the problems

$$\dot{y} = P_r(c), \quad c \in [0, 1], \quad \text{with} \quad y(0) = 0, \quad r = 0, \ldots, s-1, \qquad (1.103)$$

are solved exactly. Choosing a stepsize $h = 1$ and substituting $f(ch, y) = f(c, y) = P_r(c)$ in (1.98), gives

$$\gamma_j = \eta_j \sum_{i=1}^{k} b_i P_j(c_i) P_r(c_i) = \eta_j \int_0^1 P_j(c) P_r(c)\mathrm{d}c = \eta_j \delta_{rj},$$

---

[29] Appendix A gathers a number of relevant properties of Legendre polynomials.

where we are assuming that the quadrature formula $(c_i, b_i)_{i=1,\ldots,s}$ is exact for polynomials of degree up to $2s - 2$. From (1.97) we obtain

$$\dot{\sigma}(ch) = \sum_{j=0}^{s-1} P_j(c)\eta_j\delta_{rj} = \eta_r P_r(c),$$

and, comparing with (1.103), we get $\eta_r = 1$, $r = 0,\ldots,s-1$.

The use of the Legendre polynomials to derive a Runge-Kutta line integral method is reminiscent of the use of the W-transformation in connection with Gauss-Legendre collocation methods. In the next chapter we will extend the definition of the W-transformation to rectangular matrices showing that Runge-Kutta line integral methods defined along the Legendre basis may be thought of as a natural generalization of Gauss methods. For the time being, it is an easy exercise to check that formulae (1.78) identify the s-stage Gauss collocation method when $k = s$ and $(c_i, b_i)_{i=1,\ldots,s}$ is the Gauss quadrature formula.

This relationship does not stop here and, in fact, line integral methods of Legendre type inherit a number of relevant properties that have made Gauss collocation formulae among the most prominent integrators.[30] For example, it is not difficult to show that if the underlying quadrature rule $(c_i, b_i)_{i=1,\ldots,k}$ has order $p \geqslant 2s$, then method (1.78) has oder $2s$. This may be proved by checking the simplifying assumptions introduced by Butcher:[31]

$$B(p): \quad \sum_{i=1}^{k} b_i c_i^{q-1} = \frac{1}{q}, \qquad\qquad q = 1,\ldots,p,$$

$$C(s): \quad \sum_{j=1}^{k} \alpha_{ij} c_j^{q-1} = \frac{c_i^q}{q}, \qquad\qquad i = 1,\ldots,k, \quad q = 1,\ldots,s,$$

$$D(s-1): \quad \sum_{i=1}^{k} b_i c_i^{q-1} \alpha_{ij} = \frac{b_j}{q}(1 - c_j^q), \qquad j = 1,\ldots,k, \quad q = 1,\ldots,s-1.$$

We prefer not to go into further details about this analysis here, because the use of the Legendre basis opens a new, unexplored pathway to develop a stand-alone theory on line integral methods, including their connection with Gauss collocation methods. This new framework takes advantage of the powerful tools provided by Fourier-Legendre series of square integrable functions defined on the interval $[0,1]$, for which the shifted and normalized Legendre polynomials represents a complete orthonormal basis.[32] In fact, we observe

---

[30]However, symplecticity is not among these properties.

[31]We notice that $B(p)$ is true by assumption, while $C(s)$ comes from the previous computations to find the coefficients η_j. Finally, to prove $D(s-1)$, it is useful to interpret the sum as an exact integral, after substituting the expression for α_{ij} taken from (1.78).

[32]This series is also known as the Neumann expansion (see E.T. Whittaker and G.N. Watson. *A Course of Modern Analysis, 4-th Ed.* Cambridge University Press, 1950, p. 322).

that the coefficients γ_j, given in (1.98) with $\eta_j = 1$, may be interpreted as an approximation of the Fourier coefficients of the function $f(\sigma(ch))$:

$$\gamma_j = \sum_{i=1}^{k} b_i P_j(c_i) f(\sigma(c_i h)) \simeq \int_0^1 P_j(\tau) f(\sigma(\tau h)) d\tau,$$

the approximation being exact for polynomial functions f, provided k is large enough. Thus, under this latter assumption, the original problem

$$\dot{y}(ch) = f(y(ch)), \qquad c \in [0,1], \qquad y(0) = y_0,$$

is projected onto the finite dimensional subspace Π_s of polynomials of degree not greater than s, giving

$$\dot{\sigma}(ch) = \sum_{i=0}^{s-1} P_j(c) \int_0^1 P_j(\tau) f(\sigma(\tau h)) d\tau, \qquad \sigma(0) = y_0.$$

and the two differential equations may be easily related by exploiting classical results of the Fourier series theory.

The development of this framework forms the bulk of Chapter 3, which can be considered an advanced and independent in-depth analysis of line integral methods defined along the Legendre basis.

Bibliographical notes

For the description of Hamiltonian dynamics in the context of perturbation theory, we have followed the lucid discussions presented in [13] and [70], while for a historical perspective of the N-body problem and KAM theory, we refer the reader to [4, 12, 83, 109]. For an application of KAM theory to the stability analysis of symplectic Runge-Kutta methods, see [155] and references therein.

Analogies between the stability of nearly-integrable Hamiltonian systems and the center problem in complex dynamics, with particular reference to the issues raised by the presence of small divisors, have been discussed in [3].

Symplectic methods can be found in early work of Gröbner (see, e.g., [102]), in the pioneering work of de Vogelaere in the 50s [167], and in Ruth [151], Channell [72], Menyuk [142].

Symplectic Runge-Kutta methods have been then studied by Feng [86], Lasagni [131], Sanz-Serna [152], and Suris [162]. Such methods are obtained by imposing that the discrete map, associated with a given numerical method, is symplectic, as is the continuous one. In particular, in [152] the existence of symplectic methods of arbitrarily high order (i.e., the Gauss-Legendre collocation methods) has been established. The composition methods derived from the Störmer-Verlet formula are from [105] and based on [164]. The W-transformation has been introduced in [106] as a powerful tool to investigate algebraic stability (see also [107]).

Gauss-Legendre methods with arbitrary stages were introduced and studied by Butcher [60] (see also Ceschino and Kuntzmann [71], and Butcher [62]). A recent efficient implementation technique of such formulae has been described in [26].

Non-existence results concerning symplectic, energy-preserving formulae were stated by Ge and Marsden [98], and Chartier, Faou and Murua [73].

The conservation of the Hamiltonian over exponentially long-times has been studied by Benettin and Giorgilli in [14]. The linear growth of the global error in the solutions produced by symplectic methods applied to integrable systems has been shown in [64].

The s-stage trapezoidal methods have been used for numerically solving polynomial Hamiltonian systems in [118]; related methods can be found in [68], and generalizations are in [119, 120].

The simplifying assumptions to check the order of a Runge-Kutta method were introduced by Butcher in [60]. A proof of the order of Runge-Kutta line integral methods based on Legendre polynomials may be found in [40] and [37] (see also [48]).

Chapter 2

Examples of Hamiltonian problems

In this chapter we provide some examples of Hamiltonian problems which will be used in the sequel to test the methods. The problems are taken from a variety of real-life models, showing the richness of the applications and, therefore, supporting the use of specific methods for their solution.

2.1 Nonlinear pendulum

A classic and extensively studied nonlinear problem is the dynamics of a pendulum under influence of gravity. Using Lagrangian mechanics, it may be shown that the motion of a pendulum can be described by the dimensionless nonlinear equation

$$\ddot{q} + \sin q = 0, \tag{2.1}$$

where gravity points "downwards" and q is the angle that the pendulum forms with its stable rest position. Problem (2.1) is clearly Hamiltonian, with Hamiltonian function

$$H(q, p) = \frac{1}{2}p^2 - \cos q, \tag{2.2}$$

where $p = \dot{q}$ is the angular velocity. The state of the pendulum is completely specified by the pair $(q, p) \in [-\pi, \pi) \times \mathbb{R}$, thus a natural choice would be to consider the cylinder $(\mathbb{R}/(2\pi\mathbb{Z}) - \pi) \times \mathbb{R}$ as the phase space of the pendulum. However, it is also customary to represent the trajectories on the q-p plane rather than on a cylinder, and this can be done by expanding the cylindrical phase space by periodicity onto a phase plane.

The shape of the level curves of (2.2) in the q-p plane depends on the value of $H_0 := H(q_0, p_0)$ (see Figure 2.1). First of all we notice that values of H_0 smaller than -1 are not allowed since, otherwise, the solutions of the equation

$$\frac{1}{2}p^2 - \cos q = H_0 \tag{2.3}$$

would not be real. The following cases may arise:[1]

---

[1] We confine our study to the strip $[-\pi, \pi) \times \mathbb{R}$, and then extend the results on the phase plane, as was mentioned earlier.

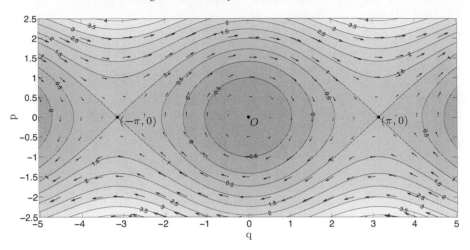

FIGURE 2.1: Level curves of (2.2). A family of periodic orbits, parametrized by the energy values $H_0 \in (-1, 1)$, originates from the stable equilibrium point O and approaches the homoclinic orbit as $H_0 \to 1^-$. The straight-up position of the pendulum is identified by the saddle points $(k\pi, 0)$, $k \in \mathbb{Z}$, which implies that the equilibrium point is unstable. Outside the homoclinic orbit, the energy is high enough (actually $H_0 > 1$) that the pendulum undertakes a rotational movement: the corresponding level curves are open and extend to infinity.

- When $H_0 = -1$, the only solution of (2.3) is $(q, p) \equiv (0, 0)$ and corresponds to the pendulum staying at rest in its stable equilibrium position.

- For $-1 < H_0 < 1$, the equation $H(q, p) = H_0$ gives rise to a closed curve, and the solution $y(t) := (q(t), p(t))$ is periodic with period $T = T(H_0)$ which is an increasing function of H_0 that tends to infinity as $H_0 \to 1^-$. These solutions correspond to oscillations of the pendulum around the straight-down stationary position (*librations*).

- For $H_0 = 1$, the trajectories tend to $(\pm\pi, 0)$ as $t \to \infty$, which correspond to the unstable equilibrium configuration with the pendulum in upright position. In fact, in this specific case, (2.3) becomes

$$\frac{1}{2}\dot{q}^2 = 1 + \cos q = 2\cos^2\frac{q}{2} \implies \dot{z} = \pm\cos z, \quad \text{with } z := \frac{q}{2}.$$

The latter is a separable first order ordinary differential equation ($\frac{dz}{\cos z} = \pm dt$) and may be solved by quadrature. The solutions read

$$\sec z + \tan z = e^{\pm(t+t_0)}.$$

This equation may be solved for z, yielding $z = \pm \arctan(\sinh(t + t_0))$, and the solution of (2.1) becomes

$$(q(t), p(t)) = \pm 2 \left(\arctan(\sinh(t + t_0)), \operatorname{sech}(t + t_0) \right). \qquad (2.4)$$

Choosing in (2.4) the $+$ sign and the $-$ sign we get, respectively,

$$\lim_{t \to \pm \infty} (q(t), p(t)) = (\pm \pi, 0), \qquad \text{and} \qquad \lim_{t \to \pm \infty} (q(t), p(t)) = (\mp \pi, 0)$$

Since $(\pi, 0)$ and $(-\pi, 0)$ identify the very same point with the pendulum in the vertical position, we conclude that the orbit is *homoclinic*.[2]

- For $H_0 > 1$ the trajectories in the phase plane are periodic in q with period 2π, and correspond to rotating movements of the pendulum (*rotations*).

The homoclinic orbit acts as a separatrix between two regions where the orbits have different topological properties, corresponding to two different modes of motion of the pendulum. Inside the separatrix, the pendulum swings back and forth around the straight-down position whereas, outside the separatrix, it continues rotating in the same direction forever. An energy-conserving method is expected to capture the correct dynamics in a neighborhood of the separatrix whatever the stepsize used.

2.2 Cassini ovals

Given the parameters $a, b > 0$, and the two points $P_1 = (-a, 0)$ and $P_2 = (a, 0)$, the corresponding Cassini oval, with foci P_1 and P_2, is defined as

$$\mathcal{C}(a, b) := \left\{ Q = (q, p) : \| Q - P_1 \|_2^2 \cdot \| Q - P_2 \|_2^2 = b^4 \right\}.$$

This curve was first investigated by the Italian astronomer Giovanni Domenico Cassini in 1680, when he was studying the relative motions of the earth and the sun. Cassini believed that the sun traveled around the earth on one of these ovals, with the earth at one focus of the oval. After him, such curves were named Cassinian (or Cassini) ovals.

The shape of the curve depends on the ratio $r = b/a$:

- when $r > 1$, the curve is a single loop, with an oval or dog-bone shape;

- when $r = 1$, one obtains a lemniscate (i.e., a curve with the shape of ∞);

---

[2] A homoclinic orbit is a trajectory in the phase space which joins a (saddle) equilibrium point to itself.

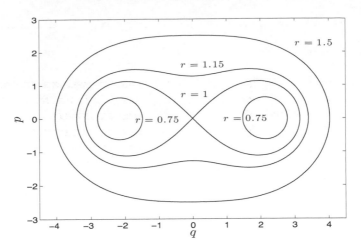

FIGURE 2.2: Cassini ovals (2.5) for various values of the ratio $r = b/a$, when $a^2 = 5$.

- when $r < 1$, the curve contains two loops surrounding the two foci.

The different cases are summarized in Figure 2.2, where the curves corresponding to the ratios $r = 1.5, 1.15, 1, 0.75$, with $a^2 = 5$, are depicted. In Cartesian coordinates, the curve is defined by the equation

$$H(q, p) := \left(q^2 + p^2\right)^2 - 2a^2\left(q^2 - p^2\right) = b^4 - a^4. \tag{2.5}$$

The function H may be regarded as a Hamiltonian function (only depending on the parameter a), whose value is kept constant along a trajectory that solves the corresponding Hamiltonian problem. More precisely, the orbit lies on the Cassini oval containing the initial condition (q_0, p_0), corresponding to the following value of the ratio $r = b/a$:

$$r = \sqrt[4]{1 + \frac{H(q_0, p_0)}{a^4}}. \tag{2.6}$$

As is the case with the pendulum problem discussed in Section 2.1, the presence of a separatrix (i.e., the lemniscate), makes the use of an energy-conserving method particularly appealing if one wants to correctly reproduce the shape of the orbit in a neighborhood of the separatrix (see Section 4.4 on page 134).

    This example suggests that an energy-conserving method may be appropriately applied to obtain a (discrete) parametrization of any regular implicit curve $H(q, p) = 0$ in the q-p plane, starting from a point (q_0, p_0) on the curve.

2.3 Hénon-Heiles problem

The Hénon-Heiles equation originates from a problem in Celestial Mechanics describing the motion of a star under the action of a gravitational potential of a galaxy which is assumed time-independent and with an axis of symmetry (the z-axis). The main question related to this model was to state the existence of a third first integral, besides the total energy and the angular momentum.[3] By exploiting the symmetry of the system and the conservation of the angular momentum, Hénon and Heiles reduced from three (cylindrical coordinates) to two (planar coordinates) the degrees of freedom, thus showing that the problem was equivalent to the study of the motion of a particle in a plane subject to an arbitrary potential $U(q_1, q_2)$:

$$H(q_1, q_2, p_1, p_2) = \frac{1}{2} \left(p_1^2 + p_2^2 \right) + U(q_1, q_2), \tag{2.7}$$

Since U in (2.7) has no symmetry in general, we cannot consider the angular momentum as an invariant anymore, so that the only known first integral is the total energy, and the question is whether or not a second integral could exist. Hénon and Heiles conducted a series of tests with the aim of giving a numerical evidence of the existence of such an integral for moderate values of the energy H, and of the appearance of chaotic behavior when H becomes larger than a critical value. In particular, for their experiments they chose

$$U(q_1, q_2) = \frac{1}{2} \left(q_1^2 + q_2^2 \right) + q_2 \left(q_1^2 - \frac{q_2^2}{3} \right), \tag{2.8}$$

which makes the Hamiltonian function a polynomial of degree three. When $U(q_1, q_2)$ approaches the value $\frac{1}{6}$, the level curves of U tends to three straight lines forming an equilateral triangle, whose vertices are saddle points of U (see Figure 2.3). These vertices have coordinates

$$Q_1 = (0, 1)^\top, \qquad Q_2 = \left(-\frac{\sqrt{3}}{2}, -\frac{1}{2} \right)^\top, \qquad Q_3 = \left(\frac{\sqrt{3}}{2}, -\frac{1}{2} \right)^\top,$$

and the triangle defines the boundary of the region

$$\mathcal{U} := \left\{ (q_1, q_2)^\top \in \mathbb{R}^2 : U(q_1, q_2) \leqslant \frac{1}{6} \right\}.$$

Consequently, by taking into account (2.7)-(2.8), one concludes that each trajectory starting at a point $\begin{pmatrix} q_1^0, & q_2^0, & p_1^0, & p_2^0 \end{pmatrix}^\top$ such that

$$H(q_1^0, q_2^0, p_1^0, p_2^0) < \frac{1}{6},$$

---

[3]Starting results about this problem were stated by G. Contopoulos [78, 79], who determined potentials such that this first integral does exist.

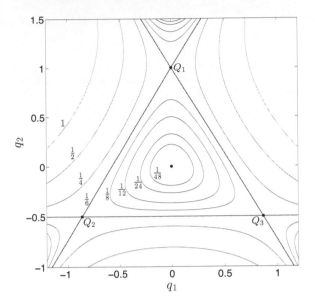

FIGURE 2.3: Level curves of Hénon-Heiles potential (2.8).

has to be contained in $\mathcal{U}$ forever. In fact, for all $t \geqslant 0$, one has:

$$U(q_1(t), q_2(t)) \leqslant H(q_1(t), q_2(t), p_1(t), p_2(t)) = H(q_1^0, q_2^0, p_1^0, p_2^0) < \frac{1}{6}.$$

In particular, it can be seen that the trajectory generates a chaotic motion for values of the energy close to $\frac{1}{6}$.[4]

2.4   N-body problem

The N-body problem concerns the motion in $\mathbb{R}^3$ of N point masses m_i, under the influence of their mutual gravitational interaction. Denoting by q_i, $i = 1, \ldots, N$, their positions in an inertial reference frame, from Newton's second law, one has that the dynamics is determined by the equations

$$m_i \ddot{q}_i = F_i, \qquad i = 1, \ldots, N, \tag{2.9}$$

---

[4]With the aid of numerical illustrations, Hénon and Heiles argued that, for example, when $H = \frac{1}{12}$ a further integral does exist while higher energy values, such as $H = \frac{1}{8}$, give rise to chaotic motion.

where F_i is the resultant of the forces acting on the i-th point mass. On the other hand, from the Newton gravitational law, this latter force equals

$$F_i = -G \sum_{j \neq i} \frac{m_i m_j (q_i - q_j)}{\|q_i - q_j\|_2^3}, \tag{2.10}$$

with $G \approx 6.67384 \cdot 10^{-11}$ [N(m/kg)^2] the gravitational constant. By setting the momenta $p_i = m_i \dot{q}_i$, and the block vectors

$$\boldsymbol{q} = \begin{pmatrix} q_1 \\ \vdots \\ q_N \end{pmatrix}, \quad \boldsymbol{p} = \begin{pmatrix} p_1 \\ \vdots \\ p_N \end{pmatrix} \in \mathbb{R}^{3N}, \tag{2.11}$$

the equations (2.9)-(2.10) turn out to be in Hamiltonian form, with Hamiltonian

$$H(\boldsymbol{q}, \boldsymbol{p}) = \frac{1}{2} \boldsymbol{p}^\top M^{-1} \boldsymbol{p} + U(\boldsymbol{q}), \tag{2.12}$$

where

$$M = \begin{pmatrix} m_1 & & \\ & \ddots & \\ & & m_N \end{pmatrix}, \quad U(\boldsymbol{q}) = -G \sum_{1 \leqslant i < j \leqslant N} \frac{m_i m_j}{\|q_i - q_j\|_2}.$$

Besides the Hamiltonian function, the other first integrals may be obtained by exploiting the symmetry of the mutual force law in (2.9)-(2.10). To this end, we begin with summing (2.9) over all particles:

$$\sum_{i=1}^N m_i \ddot{q}_i = -G \sum_{i=1}^N \sum_{j \neq i} \frac{m_i m_j (q_i - q_j)}{\|q_i - q_j\|_2^3} = 0, \tag{2.13}$$

because to each term $\frac{m_i m_j (q_i - q_j)}{\|q_i - q_j\|_2^3}$ in the latter sum, there corresponds an antisymmetric term $\frac{m_j m_i (q_j - q_i)}{\|q_j - q_i\|_2^3}$. From (2.13) we get the following nine first integrals.

- The (total) *linear momentum.* By integrating (2.13) with respect to time, we see that the sum of momenta p_i must remain constant during the motion of the particles:

$$\sum_i p_i(t) = \sum_i p_i(0) =: P_0. \tag{2.14}$$

Notice that we can always perform a linear change of coordinates so that $P_0 = 0$. Without loss of generality, we assume that this condition be satisfied in the sequel.

- The *center of mass*. By integrating (2.14) with respect to time (under condition that $P_0 = 0$), we realize that the position of the center of mass must remain constant during the motion of the particles:

$$\sum_i m_i q_i(t) = \sum_i m_i q_i(0) =: C_0. \qquad (2.15)$$

Again, a linear change of coordinates may be employed so that $C_0 = 0$ in the new inertial frame.

- The *angular momentum*. Taking in (2.9)-(2.10) the cross product by the vector q_i and summing over all particles yield

$$\sum_{i=1}^{N} m_i q_i \times \ddot{q}_i = -G \sum_{i=1}^{N} \sum_{j \neq i} \frac{m_i m_j q_i \times (q_i - q_j)}{\|q_i - q_j\|_2^3}.$$

Now $q_i \times q_i = 0$, while to each term $\frac{m_i m_j q_i \times q_j}{\|q_i - q_j\|_2^3}$ there corresponds a term $\frac{m_j m_i q_j \times q_i}{\|q_j - q_i\|_2^3}$ which is opposite in sign due to the anticommutativity property of the cross product. Thus we get

$$0 = \sum_{i=1}^{N} m_i q_i \times \ddot{q}_i = \sum_{i=1}^{N} q_i \times \dot{p}_i,$$

and hence

$$\sum_{i=1}^{N} q_i \times p_i = M_0, \qquad (2.16)$$

where M_0 is the angular momentum at the initial time.

The ten first integrals (2.12) and (2.14)–(2.16) are algebraic with respect to q_i and p_i, which means that they can be solved algebraically for one variable in terms of the others. These constants of motion allow us to reduce the number of variables of the N-body system from $6N$ to $6N - 10$. It has been a long standing objective to understand whether further first integrals could exist. However, in 1887 Bruns and later in 1896 Poincaré showed that, when $N = 3$, no new independent algebraic integral may exist. This conclusion extends to the general N-body system thus leading to the fundamental result of nonintegrability of the N-body problem for $N > 2$.

This does not mean that an analytic solution may not be found, but that it cannot be determined by transforming the problem to action-angle variables (see Section 1.3.2). Indeed, a prize was established in late 1885 posing the question of finding a uniformly convergent power series solution of the N-body problem.[5] It was awarded to the 35-year-old Poincaré not for resolving

---

[5]The prize, announced in French and German in *Acta Mathematica* and in English in *Nature*, was in honor of King Oscar II of Sweden and Norway to be awarded three years later in occasion of his 60-th birthday. The commission was composed of Mittag-Leffler, Hermite, and Weierstrass.

the question but for his remarkable contribution to the understanding of the problem.

The first successful attempt to solve the problem came several years later in 1912, when the mathematical astronomer Karl Sundman found, for the 3-body problem, a series solution in powers of $t^{1/3}$ that was convergent for any real t and for any initial conditions except for a negligible set corresponding to a vanishing angular momentum. In 1991 the Chinese student Quidong Wang extended Sundman's result to the N-body problem, though he omitted to handle the case of collisions. Unfortunately, despite that the Sundman and Wang series solutions are convergent on the whole real axis, their rate of convergence is so slow to make them useless both for qualitative investigations and for practical computations, so a numerical approach based on the use of proper geometric integrators is currently unavoidable.

2.5 Kepler problem

When the only force acting on a particle P with mass m is directed towards a fixed point, the motion is called *a central force motion*. It is easy to check that the motion of a particle in a central force field is planar. In fact, denoting by q its position and by $p = m\dot{q}$ its momentum, by definition, the force acting on P will be of the form $F = \alpha q$. Thus the angular momentum $M = q \times p$ satisfies

$$\frac{\mathrm{d}}{\mathrm{d}t} M(t) = \frac{\mathrm{d}}{\mathrm{d}t} (q(t) \times p(t)) = \dot{q}(t) \times p(t) + q(t) \times \dot{p}(t) = 0,$$

since $\dot{q}(t) \times p(t) = m\dot{q}(t) \times \dot{q}(t) = 0$ and, analogously,

$$q(t) \times \dot{p}(t) = q(t) \times m\ddot{q}(t) = q(t) \times F(t) = \alpha q(t) \times q(t) = 0.$$

Therefore, the angular momentum M remains constant with respect to time, so q and $\dot{q}$ must lie on a plane orthogonal to M, that is assumed as the coordinate reference system. Consequently, the motion of the particle is completely described by four variables:

$$q = (q_1, q_2)^\top, \quad p = (p_1, p_2)^\top \in \mathbb{R}^2,$$

yielding the evolution in time of the position and the velocity of the particle.

The Kepler problem is an inverse square central force problem because, in compliance with Newton's gravitational law, it assumes that the central force is proportional to the inverse of the square of the distance from the origin. It is epitomized by the equations

$$\ddot{q} = -\frac{q}{\|q\|_2^3}, \quad \text{with } q \in \mathbb{R}^2, \tag{2.17}$$

which form a Hamiltonian system with Hamiltonian function

$$H(q,p) = \frac{1}{2}\|p\|_2^2 - \frac{1}{\|q\|_2}. \tag{2.18}$$

The model (2.17)-(2.18) is also referred to as the 1-body problem and the motion it describes is relevant when studying the orbital movement of planets and satellites.[6] In fact, the Kepler problem may be thought of as a simplified version of the 2-body problem, obtained by introducing a change of coordinates that transforms the original system to two (symmetric) 1-body problems, where each body is immersed in an inverse-square central force field. The simplest way to transform a genuine 2-body problem

$$\ddot{q}_1 = -G\frac{m_2(q_1 - q_2)}{\|q_1 - q_2\|_2^3}, \qquad \ddot{q}_2 = -G\frac{m_1(q_2 - q_1)}{\|q_2 - q_1\|_2^3},$$

into the form (2.17) is to choose a particle, say the one with mass m_2, at the origin of the coordinate reference system, so the other particle will be located at point $q = q_1 - q_2$, and will be subject to a inverse square central force.[7] After this change of coordinates we get $\ddot{q} = -G(m_1 + m_2)\frac{q}{\|q\|_2^3}$ which is equivalent to (2.17) after the scaling $q \mapsto (G(m_1 + m_2))^{1/3}q$. We notice that, in so doing, we will have information of the position of one particle with respect to the other. If we wish to recover the positions of the particles with respect the original frame, we could exploit the property that the motion of the center of mass C, located at

$$q_C = \frac{m_1 q_1 + m_2 q_2}{m_1 + m_2}, \tag{2.19}$$

is uniform rectilinear ($\ddot{q}_C = 0$) and can be determined trivially from the initial conditions. Therefore, from the relations $q = q_1 - q_2$ and (2.19) we could express q_1 and q_2 in terms of the new variables q and q_C as (see Figure 2.4):

$$q_1 = q_C + \frac{m_2}{m_1 + m_2}q,$$

$$q_2 = q_C - \frac{m_1}{m_1 + m_2}q.$$

Since, as a Hamiltonian system, the Kepler problem admits two independent first integrals (the total energy H and the angular momentum M), it will be completely integrable and an analytic solution may be conveniently derived by considering polar coordinates (r, θ). Denoting by $e_r := (\cos\theta, \sin\theta)^\top$ and $e_\theta := (-\sin\theta, \cos\theta)^\top$ the radial and angular unit vectors, we can cast equation

---

[6]Kepler, who lived before Newton was born, postulated the motion described by (2.17) through his well-known laws, that he deduced by a direct observation of the motion of the planets around the sun.

[7]As was noted before, we always assume that the reference frame lies on the orbit plane.

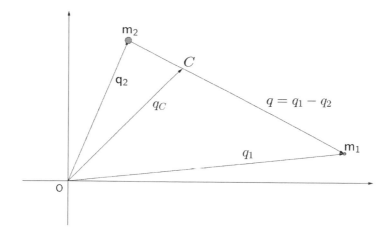

FIGURE 2.4: A visualization of the position vectors of the two particles and their center of mass C.

(2.17) as

$$\ddot{q} = -\frac{1}{\|q\|_2^2}e_r. \tag{2.20}$$

From $q(t) = r(t)e_r(t)$, and considering that $\dot{e}_r = e_\theta\dot{\theta}$, $\dot{e}_\theta = -e_r\dot{\theta}$, we get

$$\dot{q} = \dot{r}e_r + r\dot{\theta}e_\theta,$$

$$\ddot{q} = (\ddot{r} - r\dot{\theta}^2)e_r + (r\ddot{\theta} + 2\dot{r}\dot{\theta})e_\theta.$$

Hence (2.20) becomes

$$\left(\ddot{r} - r\dot{\theta}^2\right)e_r + \left(r\ddot{\theta} + 2\dot{r}\dot{\theta}\right)e_\theta = -\frac{1}{r^2}e_r, \tag{2.21}$$

while the angular momentum M is

$$M := q \times \dot{q} = (re_r) \times \left(\dot{r}e_r + r\dot{\theta}e_\theta\right)$$

$$= r^2\dot{\theta}\, e_r \times e_\theta. \tag{2.22}$$

Since M is a constant of motion, we see that its magnitude $r^2(t)\dot{\theta}(t)$ will be independent of t as well:

$$r^2(t)\dot{\theta}(t) = r^2(0)\dot{\theta}(0) =: M_0, \qquad \forall t \geqslant 0, \tag{2.23}$$

where the constant M_0 will be determined by the initial conditions. Equation (2.21) splits in two separate differential equations:

$$r\ddot{\theta} + 2\dot{r}\dot{\theta} = 0, \qquad \text{(angular component)} \tag{2.24}$$

$$\ddot{r} - r\dot{\theta}^2 = -\frac{1}{r^2}, \qquad \text{(radial component)} \tag{2.25}$$

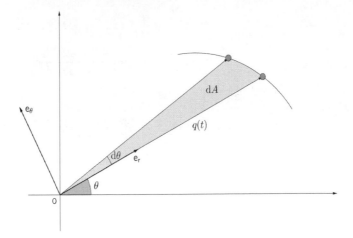

FIGURE 2.5: The rotating frame (e_r, e_θ) and the area swept by the vector $q(t)$ in the time interval $[t, t + dt]$.

that we discuss separately.

Equation (2.24) is nothing but the conservation of the angular momentum that we have discussed before (see (2.22)-(2.23)):

$$0 = r\ddot{\theta} + 2\dot{r}\dot{\theta} = \frac{1}{r}\left(r^2\ddot{\theta} + 2r\dot{r}\dot{\theta}\right) = \frac{1}{r}\frac{d}{dt}\left(r^2\dot{\theta}\right).$$

It represents the mathematical description of Kepler's second law:

The line joining the origin to the vector $q(t)$ sweeps out equal areas in equal intervals of time.

In fact, the area dA swept by $q(t) = r(t)e_r(t)$ in a time dt is $dA = \frac{1}{2}r^2d\theta$ (see Figure 2.5) and, hence, from (2.23) one has:

$$\frac{dA}{dt} = \frac{1}{2}r^2\dot{\theta} = \frac{1}{2}M_0.$$

We now consider the radial component (2.25) that, in view of (2.24), may be written as

$$\ddot{r} - \frac{M_0^2}{r^3} = -\frac{1}{r^2}, \tag{2.26}$$

where all references to the variable θ have disappeared. A further progress may be achieved by regarding the variable r as a function of θ rather than t.[8]

---

[8]Hereafter we assume that $M_0 \neq 0$. If the initial angular momentum is zero, the problem simplifies and may be solved by an elementary computation.

Exploiting again (2.24) we get, by considering that $r^2\dot\theta = M_0$,

$$\dot r = -r^2 \frac{d}{dt}\left(\frac{1}{r}\right) = -\frac{M_0}{\dot\theta}\frac{d}{dt}\left(\frac{1}{r}\right) = -M_0 \frac{d}{d\theta}\left(\frac{1}{r}\right),$$

and differentiating,

$$\ddot r = -M_0 \frac{d^2}{d\theta^2}\left(\frac{1}{r}\right)\dot\theta = -\frac{M_0^2}{r^2}\frac{d^2}{d\theta^2}\left(\frac{1}{r}\right).$$

Inserting the latter expression into (2.26) yields

$$\frac{d^2}{d\theta^2}\left(\frac{1}{r}\right) + \frac{1}{r} = \frac{1}{M_0^2}.$$

We have obtained a second order linear differential equation in the unknown $\frac{1}{r}$ whose general solution is

$$\frac{1}{r} = \frac{1}{M_0^2}\left(1 + e\cos(\theta - \theta_0)\right),$$

or

$$r(\theta) = \frac{M_0^2}{(1 + e\cos(\theta - \theta_0))}, \tag{2.27}$$

where e and θ_0 are the two constants of integration. This is the general formula in polar coordinates of a conic section that has one focus at the origin. The constant e is the eccentricity while θ_0 is the phase offset and determines the orientation of the conic in the q_1-q_2 plane: more precisely, the principal axis of the conic will form an angle θ_0 with the positive q_1-axis.[9]

Due to the symmetry of the cosine function, in describing the geometric properties of the orbit we may assume, without loss of generality, e non negative. The four possibilities are summarized in the list below:

$e = 0$: the orbit is a circle (of radius M_0^2);

$e < 1$: the orbit is an ellipse (with semi-major axis M_0^2);

$e = 1$: the orbit is a parabola;

$e > 1$: the orbit is a hyperbola.

In particular, solutions starting at

$$(q, p)^\top := \left(1 - e, \quad 0, \quad 0, \quad \sqrt{\frac{1+e}{1-e}}\ \right)^\top, \quad \text{with} \quad e \in [0, 1), \tag{2.28}$$

---

[9]The principal axis of a conic is the line passing through the focus and perpendicular to the directrix. For ellipses it will include the major axis and is also called line of apsides.

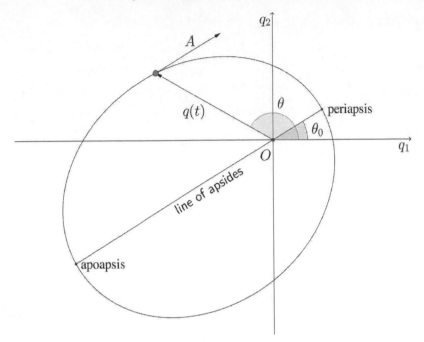

FIGURE 2.6: An elliptic Kepler orbit as described by equation (2.27). The Lenz vector $A(q,p)$ is also highlighted.

may be shown to be periodic of period $T = 2\pi$, and describe an ellipse of eccentricity e in the q_1-q_2 plane.

Besides the Hamiltonian function (2.18) and the angular momentum [10]

$$M(q,p) := q_1 p_2 - p_1 q_2, \qquad (2.29)$$

an inverse-square central force field gives birth to an additional invariant, called Laplace-Runge-Lenz (LRL or Lenz) vector. In order to properly introduce this vector, we consider again the general setting of a Kepler problem defined in the configuration space $\mathbb{R}^3$, so $q = (q_1, q_2, q_3)^\top$ and $p = (p_1, p_2, p_3)^\top$ for a while. The Lenz vector then takes the form

$$A(q,p) := p \times (q \times p) - \frac{q}{\|q\|_2},$$

and is evidently orthogonal to the angular momentum $q \times p$ so it lies on the orbital plane. A direct computation may be employed to prove that $A(q(t), p(t))$

---

[10]Since the reference plane on which the dynamics takes place has been chosen orthogonal to the angular momentum (see the beginning of the section), we may think of this latter physical quantity as a scalar function, namely the third component of the cross product $(q_1, q_2, q_3) \times (p_1, p_2, p_3)$.

is a constant of motion. In fact, on the one hand we have

$$
\frac{\mathrm{d}}{\mathrm{d}t} p \times (q \times p) = \ddot{q} \times (q \times p) + p \times \frac{\mathrm{d}}{\mathrm{d}t}(q \times p) = \ddot{q} \times (q \times p)
$$

$$
= -\frac{q}{\|q\|_2^3} \times (q \times p) = -\frac{1}{\|q\|_2^3}\left((q^\top p)q - \|q\|_2^2 p\right),
$$

where we have used the property

$$
u \times (v \times w) = (u^\top w)v - (u^\top v)w, \qquad \text{for all} \ \ u, v, w \in \mathbb{R}^3.
$$

On the other hand,

$$
\frac{\mathrm{d}}{\mathrm{d}t}\left(\frac{q}{\|q\|_2}\right) = \frac{p}{\|q\|_2} - \frac{q^\top p}{\|q\|_2^3} q,
$$

so that $\frac{\mathrm{d}}{\mathrm{d}t} A(q(t), p(t)) = 0$. Choosing the orbit plane as the reference plane, according to what was done throughout the present section, we see that

$$
A(q, p) = \begin{pmatrix} A_1(q, p) \\ A_2(q, p) \\ A_3(q, p) \end{pmatrix} \equiv \begin{pmatrix} p_2 M(q, p) - q_1/\|q\|_2 \\ -p_1 M(q, p) - q_2/\|q\|_2 \\ 0 \end{pmatrix}, \qquad (2.30)
$$

where the angular momentum $M(q, p)$ is defined in (2.29).

It is easy to check that the four constants of motion $H(q, p)$, $M(q, p)$, $A_1(q, p)$ and $A_2(q, p)$ are related by the following equation

$$
A_1^2(q, p) + A_2^2(q, p) = 1 + 2H(q, p)M^2(q, p), \qquad (2.31)
$$

so at most three of them may be independent (for example H, M and A_2). This is consistent with the general rule that a Hamiltonian system with m degrees of freedom may admit at most $2m - 1$ independent first integrals. A system with more than m constants of motion is called *super-integrable* and a system with $2m - 1$ constants of motion, as is the Kepler problem, is called *maximally super-integrable*. Since the orbit lies in the intersection in the phase-space of the manifolds representing the first integrals, a consequence of the maximal super-integrability of the Kepler problem is the existence of closed periodic orbits for a wide range of initial data.

In a Kepler problem, the Lenz vector can be used to describe the shape and orientation of the orbit of one particle around the other. To see this, we consider the inner product of the two vectors A and q and denote by φ the angle they form. On the one hand we have (see (2.31))

$$
A^\top q = \|A\|_2 \|q\|_2 \cos\varphi = \sqrt{1 + 2HM^2} \, \|q\|_2 \cos\varphi, \qquad (2.32)
$$

and, on the other hand, from (2.29) and (2.30) we obtain

$$
A^\top q = A_1 q_1 + A_2 q_2 = \left(p_2 M - \frac{q_1}{\|q\|_2}\right) q_1 - \left(p_1 M + \frac{q_2}{\|q\|_2}\right) q_2 = M^2 - \|q\|_2.
$$

Equaling this latter expression with the rightmost term in (2.32) yields

$$\|q\|_2 \; = \; \frac{M^2}{\left(1 + \sqrt{1 + 2HM^2}\cos(\varphi)\right)}. \qquad (2.33)$$

The conservation of the Lenz vector thus provides an alternative tool to derive the equation of the Kepler orbit. Comparing (2.33) with (2.27) we deduce the following properties:

1. $\varphi = \theta - \theta_0$, thus the Lenz vector is parallel to the principal axis and points towards the periapsis (see Figure 2.6).

2. The eccentricity e may be expressed in terms of the total energy and the angular momentum as

$$e^2 \; = \; 1 + 2HM^2 \; = \; \|A\|_2^2,$$

where the last equality follows from (2.31). Consequently, to a negative energy H there corresponds an elliptic orbit since $e \in (-1, 1)$, when the energy is precisely zero then $e = 1$ and the orbit is a parabola, and finally, if the energy is positive, $|e| > 1$ and we get a hyperbola.

The fact that $e = \|A\|_2$ justifies the name of *eccentricity vector* often used to denote the Lenz vector $A(q, p)$ of the normalized Kepler problem (2.17).

2.6 Circular restricted three-body problem

The approximately circular motion of the planets around the sun and the small masses of the asteroids and satellites compared to planetary masses suggested the formulation of the *circular restricted three-body problem* (CRTBP). It models the motion of a body with negligible mass (planetoid) in the gravitational field generated by two celestial bodies with finite mass (primaries) rotating around their common center of mass in circular orbits. It is therefore a simplified instance of the three-body problem, where the term *restricted* reflects the presence of a massless body that does not affect the motion of the other two bodies, while the term *circular* means that the two massive bodies revolve in the same plane around their common center of mass C_M on circular orbits (with the same angular velocity).[11]

Its interest goes back to the second quarter of the eighteenth century, in the context of the lunar theory.[12] A renewed interest arose starting from the late 1960s up to present day and is testified by a rich and growing literature

---

[11]Hereafter, we shall denote the center of mass by C_M, since C will be used later to denote the Jacobi constant.

[12]It was first introduced by Euler in 1767 who studied the dynamics and determined

on the design and analysis of a variety of orbits connected with the motion of spacecrafts, satellites and asteroids (see the bibliographical notes).

Discarding the effect of gravitational attraction of the smallest body implies that the three-body problem splits in a two-body problem governing the planar motion of the two primaries, as described in Section 2.5, plus a one-body problem describing the motion of the planetoid subject to the gravitational attraction of the primaries. We notice that these simplifying assumptions imply that Newton's third law is violated because, if the primaries affect the motion of the planetoid, this latter should also influence the motion of the primaries. As a consequence, the total energy of the CRTBP is no longer strictly conserved. However, as we will see, the problem still admits a Hamiltonian function that is conserved.

Without loss of generality, we will illustrate the problem by considering the concrete case where the two primaries are the sun and the earth+moon whose masses are denoted m_1 and m_2 respectively.[13] Usually the units are normalized and chosen so that the properties of the resulting dynamical system only depend on the single parameter

$$\mu := \frac{m_2}{m_1 + m_2}.$$

In our situation we have $\mu = 3.04036 \cdot 10^{-6}$. To obtain dimensionless coordinates the following normalizing assumptions are introduced:

1. the total mass of the system is $m_1 + m_2 = 1$ (thus $m_2 = \mu$ and $m_1 = 1 - \mu$);

2. the unit of length is the distance between the two primaries (in our case $R = 1.49589 \cdot 10^8$km);

3. the unit of time is $1/n$, where n is the constant angular velocity of the primaries around their center of mass C_M. Consequently the (normalized) revolution time will be $T = 2\pi$ (for the sun–earth+moon system $n = 1.99099 \cdot 10^{-7}$rad/s).

Notice that, from the above hypotheses, the gravitational constant is unity, $G = 1$, while the distances of the bodies with mass m_1 and m_2 from the center of mass are $1 - \mu$ and μ, respectively (see (2.19)).

Considering that the motion of the two primaries is completely known, we see that the CRTBP only requires six variables for its description, namely the three (generalized) coordinates and the three conjugate momenta of the planetoid: in the inertial frame centered at the center of mass C_M, they will be denoted by $Q = (Q_1, Q_2, Q_3)^\top$ and $P = (P_1, P_2, P_3)^\top$, respectively. Analogously,

---

three equilibria. Few years later Lagrange (1772) determined other 2 equilibria. In 1836 Jacobi discovered the integral of the motion in the synodic (rotating) reference system. Poincaré (1899) carried out an intensive investigation to determine its stability properties and qualitative characteristics.

[13]The eccentricity of Earth's orbit is currently about 0.0167. Thus the approximation to a circular orbit turns out to be appropriate and is widely used in the literature.

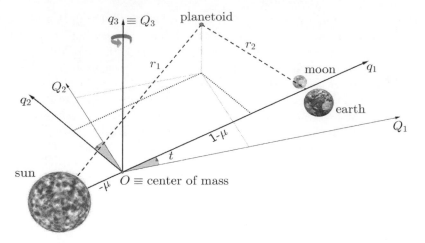

FIGURE 2.7: The circular restricted three body problem adapted to the sun–earth+moon system (for visualization purposes, dimensions and distances do not match real scales).

Q_s, $Q_{em} \in \mathbb{R}^3$ will denote the coordinates of the sun and the earth+moon in the inertial frame. It turns out that the dimensionless equations governing the motion of the planetoid read

$$\ddot{Q} = -(1-\mu)\frac{Q - Q_s(t)}{\|Q - Q_s(t)\|_2^3} - \mu\frac{Q - Q_{em}(t)}{\|Q - Q_{em}(t)\|_2^3}, \qquad (2.34)$$

where, assuming that at time zero both the sun and the center of mass of the earth+moon lie on the Q_1 axis,

$$Q_s(t) = -\mu\begin{pmatrix}\cos t \\ \sin t \\ 0\end{pmatrix}, \qquad Q_{em}(t) = (1-\mu)\begin{pmatrix}\cos t \\ \sin t \\ 0\end{pmatrix}.$$

Defining the conjugate momenta vector as the vector of velocities, $P = \dot{Q}$, we see that (2.34) is a Hamiltonian system with Hamiltonian function

$$H(Q,P,t) = \frac{1}{2}P^\top P - \frac{1-\mu}{\|Q - Q_s(t)\|_2} - \frac{\mu}{\|Q - Q_{em}(t)\|_2}.$$

The time dependence of the sun and the earth+moon coordinates is responsible for the non autonomous character of the system. To get rid of this explicit time dependance, it is common to write down the equations of motion of the planetoid in a coordinate system $Oq_1q_2q_3$ where the primaries are stationary. This is accomplished by introducing a rotating (synodic) orthogonal frame centered at C_M, with the q_1-q_2 axes lying in the plane of the sun-earth/moon orbit, the q_1-axis being oriented from the sun toward the earth+moon, and the q_3-axis forming a right-hand frame with the other axes. Thus, the sun and the earth+moon are located on the q_1-axis at the abscissae $-\mu$ and $1 - \mu$, respectively (see Figure 2.7).

The change of variables relating the coordinates $Q = (Q_1, Q_2, Q_3)^\top$ of the inertial frame to $q = (q_1, q_2, q_3)^\top$ of the synodic frame are

$$Q = \Psi(t)q, \qquad \text{with} \quad \Psi(t) = \begin{pmatrix} \cos t & -\sin t & 0 \\ \sin t & \cos t & 0 \\ 0 & 0 & 1 \end{pmatrix}. \tag{2.35}$$

In the new coordinate system, the sun and the earth+moon are located at

$$q_{\text{s}} = (-\mu, 0, 0)^\top, \qquad q_{\text{em}} = (1 - \mu, 0, 0)^\top.$$

Furthermore, since the linear transformation defined by matrix $\Psi(t)$ employes a (counter-clockwise) rotation of the q_1-q_2 plane of an angle t, $\Psi(t)$ is a unitary matrix and consequently the distances of the planetoid from the sun and the earth+moon remain unchanged:

$$r_1 \;:=\; \|Q - Q_{\text{s}}\|_2 \equiv \|q - q_{\text{s}}\|_2 = ((q_1 + \mu)^2 + q_2^2 + q_3^2)^{1/2},$$
$$r_2 \;:=\; \|Q - Q_{\text{em}}\|_2 \equiv \|q - q_{\text{em}}\|_2 = ((q_1 - (1 - \mu))^2 + q_2^2 + q_3^2)^{1/2}. \tag{2.36}$$

Defining the matrices $\hat{J}$ and $\hat{I}$ as

$$\hat{J} := \begin{pmatrix} 0 & 1 & 0 \\ -1 & 0 & 0 \\ 0 & 0 & 0 \end{pmatrix}, \qquad \text{and} \quad \hat{I} := \begin{pmatrix} 1 & 0 & 0 \\ 0 & 1 & 0 \\ 0 & 0 & 0 \end{pmatrix},$$

the following identities hold true:

$$\Psi^{-1}(t)\dot{\Psi}(t) = -\hat{J}, \qquad \Psi^{-1}(t)\ddot{\Psi}(t) = -\hat{I} = \hat{J}^2.$$

Inserting (2.35) into (2.34) we get

$$\ddot{q} - 2\hat{J}\dot{q} - \hat{I}q = -(1 - \mu)\frac{q - q_{\text{s}}}{r_1^3} - \mu\frac{q - q_{\text{em}}}{r_2^3}.$$

Now we define the vector of conjugate momenta as

$$p(t) = (p_1(t), p_2(t), p_3(t))^\top := (\dot{q}_1(t) - q_2(t), \dot{q}_2(t) + q_1(t), \dot{q}_3(t))^\top \equiv \dot{q}(t) - \hat{J}q(t).$$

The system becomes, by taking into account (2.36):

$$\dot{q} = p + \hat{J}q, \tag{2.37}$$
$$\dot{p} = \hat{J}p - (1 - \mu)\frac{q - q_{\text{s}}}{r_1^3} - \mu\frac{q - q_{\text{em}}}{r_2^3}. \tag{2.38}$$

In conclusion, the Hamiltonian function in non-dimensional form associated with the dynamical system governing the motion of the planetoid in the synodic frame is

$$H(q, p) = p_1 q_2 - p_2 q_1 + \frac{1}{2}p^\top p - \frac{1 - \mu}{r_1} - \frac{\mu}{r_2}, \tag{2.39}$$

and does not depend on the time explicitly.

2.6.1 Planar CRTBP

If the initial conditions for the planetoid are in the form $(q_1^{(0)}, q_2^{(0)}, 0)^\top$ $(p_1^{(0)}, p_2^{(0)}, 0)^\top$, the planetoid will move in the same plane as the primaries and the corresponding system will be described by the four (rather than six) variables $q(t) = (q_1(t), q_2(t))^\top$ and $p(t) = (p_1(t), p_2(t))^\top = (\dot{q}_1(t) - q_2(t), \dot{q}_2(t) + q_1(t))^\top$. The Hamiltonian function is again (2.39) with

$$r_1 := ((q_1 + \mu)^2 + q_2^2)^{1/2}, \quad \text{and} \quad r_2 := ((q_1 - (1 - \mu))^2 + q_2^2)^{1/2}.$$

The resulting system is referred to as the *planar restricted three-body problem* and the equations of motion are in the form (2.37)-(2.38) with $\hat{J}$ replaced with J:

$$\dot{q} = p + Jq, \tag{2.40}$$

$$\dot{p} = Jp - (1 - \mu)\frac{q - q_s}{r_1^3} - \mu\frac{q - q_{em}}{r_2^3}. \tag{2.41}$$

An in-depth inspection of the Hamiltonian function reveals its relationship with a well known first integral introduced by Jacobi in 1836, namely the *Jacobi constant* or *Jacobi integral*. We begin with substituting $p_1 = \dot{q}_1 - q_2$ and $p_2 = \dot{q}_2 + q_1$ in the first three terms of (2.39):

$$p_1 q_2 - p_2 q_1 + \frac{1}{2}(p_1^2 + p_2^2)$$

$$= (\dot{q}_1 - q_2)q_2 - (\dot{q}_2 + q_1)q_1 + \frac{1}{2}\left((\dot{q}_1 - q_2)^2 + (\dot{q}_2 + q_1)^2\right)$$

$$= \frac{1}{2}\left(\dot{q}_1^2 + \dot{q}_2^2\right) - \frac{1}{2}\left(q_1^2 + q_2^2\right).$$

Thus, the Hamiltonian function (2.39) may be equivalently expressed in terms of the positions and velocities of the planetoid in the rotating frame, which allows us to highlight the contribution to the total energy coming from the *kinetic energy* in the rotating frame, the *centrifugal potential*, and the *gravitational potential*:

$$H(q, p) = \underbrace{\frac{1}{2}\left(\dot{q}_1^2 + \dot{q}_2^2\right)}_{\substack{\text{kinetic} \\ \text{energy}}} - \underbrace{\frac{1}{2}\left(q_1^2 + q_2^2\right)}_{\substack{\text{centrifugal} \\ \text{potential}}} - \underbrace{\left(\frac{1 - \mu}{r_1} + \frac{\mu}{r_2}\right)}_{\substack{\text{gravitational} \\ \text{potential}}}. \tag{2.42}$$

The term

$$\Omega(q_1, q_2) = \frac{1}{2}\left(q_1^2 + q_2^2\right) + \frac{1 - \mu}{r_1} + \frac{\mu}{r_2}, \tag{2.43}$$

may be viewed as the *effective potential* of the system in the synodic frame and plays an important role in the determination of the equilibrium points of

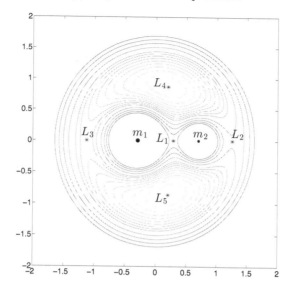

FIGURE 2.8: Level curves of $-\Omega(q_1, q_2)$ and the five Lagrangian points. In a neighborhood of each of the two primaries there is a gravity well: $\lim_{r_1 \to 0} -\Omega(q_1, q_2) = \lim_{r_2 \to 0} -\Omega(q_1, q_2) = -\infty$. Lighter-colored curves correspond to increasing values of the energy ($-\Omega(q_1, q_2)$ reaches its maximum value at L_4 and L_5).

the planar CRTBP in the synodic frame: these are the stationary points of $\Omega(q_1, q_2)$, namely the solutions of

$$\frac{\partial}{\partial q_1}\Omega(q_1, q_2) = 0, \qquad \frac{\partial}{\partial q_2}\Omega(q_1, q_2) = 0.$$

Such equations admit five solutions referred to as *Lagrangian* or *libration* points: three saddle points, L_1, L_2, L_3, which are collinear with the primaries, and two minima, L_4 and L_5, which form an equilateral triangle with them (see Figure 2.8). Periodic and quasi-periodic orbits around libration points are suited for a number of mission applications. For example, sun-earth libration points are commonly used for deep space or sun activity observations.[14]

The *Jacobi integral* is defined as

$$C = 2\Omega(q_1, q_2) - \|\dot{q}\|_2^2. \tag{2.44}$$

From (2.42) it follows that $C = -2H$, so the Jacobi integral has not to be interpreted as a further independent constant of motion.[15]

---

[14]A stability analysis shows that the collinear equilibrium points L_1, L_2 and L_3 are unstable, while L_4 and L_5 are stable for $\mu < 0.03852$. An example are the Trojan asteroids librating around the points L_4 and L_5 in the Sun-Jupiter system.

[15]In fact, according to what stated in Section 2.4, there are no other independent con-

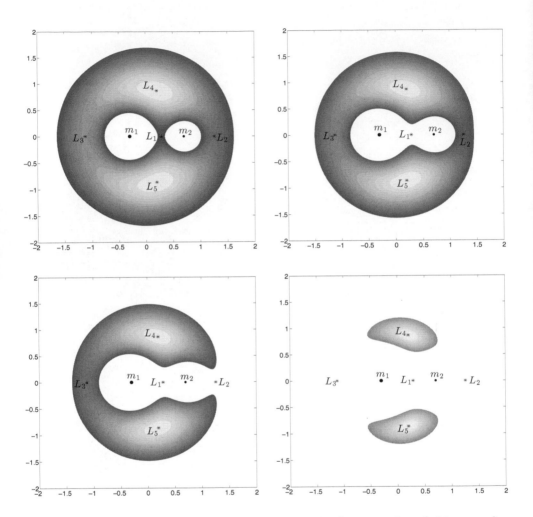

FIGURE 2.9: Possible shapes of Hill regions in the q_1-q_2 plane (white areas) for increasing values of energy. The forbidden regions (filled gray areas) shrink as the energy increases and eventually vanish at the Lagrange points L_4 and L_5. Upper left picture: $C > C_1$; upper right picture: $C_2 < C < C_1$; bottom left picture: $C_3 < C < C_2$; bottom right picture: $C_{4/5} < C < C_3$.

One interest in the Jacobi integral relies on the fact that, once the value of C is derived from the initial conditions of the planetoid, equation (2.44) may be exploited to compute the admissible region, in the configuration space, for the motion of the planetoid. In fact, since the term $\|\dot{q}\|_2^2$ is nonnegative, from (2.44) we get

$$\Omega(q_1, q_2) \geqslant \frac{C}{2} (= -H). \tag{2.45}$$

Since Ω is a function of positions only, inequality (2.45) defines a subset R_C of the q_1-q_2 plane containing all the points that could be potentially reached by the orbit as the time progresses or, equivalently, the complement of R_C defines the forbidden region for the given orbit. The sets R_C are referred to as *Hill regions*, while their boundaries are called *zero velocity curves* and correspond to the set of points (q_1, q_2) such that $2\Omega(q_1, q_2) = C$.

Let us denote by $C_i := 2\Omega(L_i)$ the Jacobi constants corresponding to the zero velocity curves passing through the Lagrangian points L_i, $i = 1, \ldots, 5$. Figure 2.9 summarizes the possible shapes of the Hill regions for $\mu = 0.3$ as the value of C decreases.[16] When $C > C_1$ (upper-left picture), the Hill region consists of three non-connected areas: two of them surrounding the primaries and the third one unbounded and far from the primaries. Consequently, the planetoid may orbit one of the two primaries without escaping or otherwise may move on a wide outer region. As C approaches the value C_1, the two inner bounded regions merge at the point L_1 and, for $C_2 < C < C_1$, the planetoid may circle around both primaries as time progresses, in a bounded region embracing both of them (upper-right picture). When $C = C_2$ a connection of the inner and outer allowed regions emerges at point L_2, and right beyond that value ($C_3 < C < C_2$), the forbidden region forms a horseshoe shape figure (bottom-left picture). The planetoid may eventually transit along the neck region in the vicinity of the Lagrange point L_2 thus escaping the system. Analogously, at $C = C_3$ the forbidden region splits in two separate areas surrounding the points L_4 and L_5 (bottom-right picture). If C is further reduced, the forbidden region shrink until they disappear at the Lagrange points L_4 and L_5, for $C = C_{4/5}$, which is the minimum value that 2Ω may assume.

2.6.2 Hill's lunar problem

The Hill problem is a special, simplified case of the planar CRTBP. It studies the motion of the planetoid in a neighborhood of the body with mass μ, under the assumption that μ is sufficiently small. It was first formulated by the American astronomer George William Hill with the aim of studying

---

served quantities besides the Hamiltonian function (the conservation of the center of mass, the linear and angular momenta have been exploited to reduce the number of degrees of freedom during the definition of the circular restricted version of the three-body problem).

[16]The approximate values of C_i when $\mu = 0.3$ are: $C_1 \approx 3.9201496$, $C_2 \approx 3.5564130$, $C_3 \approx 3.2913502$ and $C_{4/5} = 2.79$.

the motion of the moon under the influence of the gravitational effects of the earth and the sun. For this reason, in the sequel, we will call *sun* the body with mass $1 - \mu$, *earth* the body with mass μ and *moon* the planetoid. We will define the Hill problem by performing a sequence of symplectic change of variables and a scaling on the Hamiltonian function associated with the planar CRTBP (see (2.39)):

$$H(q,p) = p_1 q_2 - p_2 q_1 + \frac{1}{2}(p_1^2 + p_2^2) - \frac{1-\mu}{r_1} - \frac{\mu}{r_2},$$

with $r_1 = ((q_1 + \mu)^2 + q_2^2)^{1/2}$ and $r_2 = ((q_1 - (1 - \mu))^2 + q_2^2)^{1/2}$.

To begin with, it is convenient to choose the origin of the synodic frame at the point where the earth is located. This is accomplished by means of the change of coordinates

$$q_1 \to q_1 + (1 - \mu), \qquad q_2 \to q_2,$$

and consequently, since $p_1 = \dot{q}_1 - q_2$ and $p_2 = \dot{q}_2 + q_1$, we get

$$p_1 \to p_1, \qquad p_2 \to p_2 + (1 - \mu).$$

Here and in the sequel we will continue to denote the new variables by q_1, q_2, p_1 and p_2, and we will always discard the constant terms that do not affect the equations of motion. After this (symplectic) change of coordinates, the Hamiltonian function reads

$$H(q,p) = p_1 q_2 - p_2 q_1 + \frac{1}{2}(p_1^2 + p_2^2) - (1-\mu)q_1 - \frac{1-\mu}{((q_1 + 1)^2 + q_2^2)^{1/2}} - \frac{\mu}{(q_1^2 + q_2^2)^{1/2}}.$$

The assumption on the location of the planetoid (the moon) permits a simplification of the equations describing its dynamics. Essentially, one discards the terms of order three or higher in q_1 and q_2 in the Taylor expansion around $(0,0)$ of the term corresponding to the gravitational potential generated by the sun:

$$\frac{1-\mu}{((q_1 + 1)^2 + q_2^2)^{1/2}} \simeq (1-\mu)\left(1 - q_1 + q_1^2 - \frac{1}{2}q_2^2\right).$$

By inserting this truncated expansion in the previous equation (and neglecting the constant term), we obtain the following simplified Hamiltonian function:

$$\tilde{H}(q,p) := p_1 q_2 - p_2 q_1 + \frac{1}{2}(p_1^2 + p_2^2) - \frac{\mu}{(q_1^2 + q_2^2)^{1/2}} - (1-\mu)\left(q_1^2 - \frac{1}{2}q_2^2\right).$$

Since, by assumption, μ is very small, we can drop the factor $(1-\mu)$ in the last term, since $1 - \mu \approx 1$. To make the equations independent of the parameter μ, we now perform the following scaling

$$q_1 \to \mu^{1/3} q_1, \qquad q_2 \to \mu^{1/3} q_2,$$

which defines a symplectic transformation with $\mu^{-2/3}$ as multiplier.[17] The Hamiltonian function defining the Hill problem is defined as $K(q,p) := \mu^{-2/3}\widetilde{H}(\mu^{1/3}q, \mu^{1/3}p)$ (see footnote 17), and one can check that

$$K(q,p) = p_1 q_2 - p_2 q_1 + \frac{1}{2}(p_1^2 + p_2^2) - \frac{1}{(q_1^2 + q_2^2)^{1/2}} - q_1^2 + \frac{1}{2}q_2^2. \qquad (2.46)$$

This reduced system admits only two equilibrium points located on the q_1-axis on both sides of the earth: $L_1 = \left(-(1/3)^{1/3}, 0\right)$ and $L_2 = \left((1/3)^{1/3}, 0\right)$.

2.6.3 Optimal transfer trajectory

An interesting problem in astrodynamics is the *optimal transfer trajectory*, which consists of finding the optimal control laws that drive a spacecraft from an initial state, say P_1, to a desired final state P_2 in a given time T. Here, the term *optimal* means that the amount of propellant needed to produce the change in orbital elements is minimized.

Since the fuel consumption is proportional to changes in the velocity, an input vector $u(t) = \left(\begin{array}{ccc} u_1(t), & u_2,(t) & u_3(t) \end{array} \right)^{\top}$ enters the dynamical system to control the acceleration of the vehicle along the three axes. This is accomplished by considering a new non-autonomous Hamiltonian function,

$$\bar{H}(q,p) = H(q,p) + q^{\top}u,$$

where $H(q,p)$ is the Hamiltonian of the original problem (e.g., (2.39)). Our optimal control problem is then formulated as follows:

> *Minimize the quadratic cost $\frac{1}{2}\int_0^T \|u(t)\|_2^2 dt$, subject to the dynamics induced by $\bar{H}(q,p)$ and the boundary conditions corresponding to the states P_1 and P_2.*

We assume that the control input is unconstrained and regular. The Pontryagin maximum principle is often used to attack this problem. Setting $y^{\top} = \left(\begin{array}{cc} q^{\top}, & p^{\top} \end{array} \right)$ (state variables) and $\lambda = \left(\begin{array}{ccc} \lambda_1, & \ldots, & \lambda_6 \end{array} \right)^{\top}$ (costate variables), one considers the augmented Hamiltonian function

$$\widetilde{H}(y, \lambda, u) = \frac{1}{2}u^{\top}u + \lambda^{\top}J\nabla\bar{H}(q,p).$$

Then, the necessary conditions for optimality are

$$\dot{y} = \frac{\partial\widetilde{H}}{\partial\lambda}, \qquad \dot{\lambda} = -\frac{\partial\widetilde{H}}{\partial y}, \qquad \frac{\partial\widetilde{H}}{\partial u} = 0.$$

---

[17] A transformation $\psi : z \in \mathbb{R}^{2m} \mapsto y \in \mathbb{R}^{2m}$ is said *symplectic with multiplier* μ if its Jacobian matrix satisfies the identity $\mu\psi'(z)J\psi'(z)^{\top} = J$ for all z. For $\mu = 1$, we get the classical definition of a symplectic map. Employing such a change of variables in a Hamiltonian system $\dot{y} = J\nabla H(y)$ yields

$$\dot{z} = (\psi'(z))^{-1}J\nabla H(\psi(z)) = \mu J\psi'(z)^{\top}\nabla H(\psi(z)) = J\nabla_z(\mu H(\psi(z))),$$

that is a Hamiltonian system with the new Hamiltonian function $K(z) := \mu H(\psi(z))$.

The third equation gives $u_i = -\lambda_{(3+i)}$, $i = 1, 2, 3$, so that the resulting system is autonomous and only depends on the state and costate variables. It is defined by the Hamiltonian

$$\hat{H}(y, \lambda) = \frac{1}{2}(\lambda_4^2 + \lambda_5^2 + \lambda_6^2) + \lambda^\top \left(J\nabla H(q, p) - (0, 0, 0, \lambda_4, \lambda_5, \lambda_6)^\top \right)$$

$$= \lambda^\top J\nabla H(q, p) - \frac{1}{2}(\lambda_4^2 + \lambda_5^2 + \lambda_6^2). \tag{2.47}$$

Of course, the previous arguments also apply when the problem is planar (e.g., (2.40)-(2.41) or the one derived by the simplified Hamiltonian (2.46)). We will see an example of optimal transfer trajectory for the Hill problem in Chapter 6, for which (2.47) becomes, by setting $y = \begin{pmatrix} q_1, & q_2, & p_1, & p_2 \end{pmatrix}^\top$ and $\lambda = \begin{pmatrix} \lambda_1, & \dots, & \lambda_4 \end{pmatrix}^\top$:

$$\hat{H}(y, \lambda) = \lambda_1 p_1 + \lambda_2 p_2 + \lambda_3 \left(2p_2 - \frac{q_1}{(q_1^2 + q_2^2)^{\frac{3}{2}}} + 3q_1 - \lambda_3 \right)$$

$$+ \frac{\lambda_3^2 + \lambda_4^2}{2} - \lambda_4 \left(2p_1 + \frac{q_2}{(q_1^2 + q_2^2)^{\frac{3}{2}}} + \lambda_4 \right). \tag{2.48}$$

2.7 Fermi-Pasta-Ulam problem

In 1953 Enrico Fermi, John Pasta and Stanislaw Ulam undertook a series of numerical experiments on the Los Alamos computer *Maniac I* concerning the simulation of the motion of m particles disposed along a line and coupled by vibrating springs.[18] They assumed the forces generated by the strings to be nonlinear in nature and, in particular, they considered the cases where the forces displayed quadratic, cubic, and piecewise linear terms. Denoting by h the spring length and by $X_i(t)$ the position of the i-th point mass on the string we may write,

$$X_i(t) = ih + q_i(t), \qquad i = 1, \dots, m,$$

where ih is the rest position of the i-th particle and $q_i(t)$ is its displacement from the equilibrium. For the quadratic and cubic forces the equations of motion read, for $i = 1, \dots, m$,

$$\ddot{q}_i = (q_{i+1} - 2q_i + q_{i-1}) + \alpha \left((q_{i+1} - q_i)^2 - (q_i - q_{i-1})^2 \right) \tag{2.49}$$

and

$$\ddot{q}_i = (q_{i+1} - 2q_i + q_{i-1}) + \beta \left((q_{i+1} - q_i)^3 - (q_i - q_{i-1})^3 \right) \tag{2.50}$$

---

[18]They chose $m = 32$ and $m = 64$ in their experiments.

where, assuming fixed boundaries, $q_0 = q_{m+1} = 0$. The equations in (2.49) are referred to as the α-*model* whereas those in (2.50) form the β-*model*.

Their study was motivated by a previous theoretical investigation conducted by the physical chemist Peter Debye in 1914, aiming at explaining the finite thermal conductivity of solids. To model the heat conductivity phenomenon in solids, Debye represented the continuum by a one-dimensional lattice composed by masses connected by springs, and conjectured that the finiteness of thermal conductivity might stem from the anharmonicity of forces exerted by the springs. In fact, in presence of linear springs, the energy would be propagated, without diffusion, by the fundamental modes of the system and the thermal conductivity would be infinite. On the other hand, he thought that the presence of a weak nonlinear interaction could hinder the propagation of energy, permitting the fundamental modes (of the corresponding linearized model) to exchange energy, thus realizing the transition to thermal equilibrium.

Consistently with Debye's conjecture, Fermi, Pasta and Ulam considered an initial configuration of their nonlinear models where all the energy was concentrated in one or few modes (actually the lowest) and argued that, on the long time, the system would evolve towards a state where the energy is equally distributed among all modes on average. This behavior was referred to as *thermalization*, and the transition time from the initial state to this statistical equilibrium would have furnished information about the diffusion coefficient.

The simplest initial condition they considered was the one with all the energy concentrated within the lowest mode. Contrary to their expectations, the behavior they could experience was far from being an eventual equidistribution of the energy among the different modes of motion:

> *Let us say here that the results of our computations show features which were, from the beginning, surprising to us. Instead of a gradual continuous flow of energy from the first mode to the higher modes, all of the problems show an entirely different behavior. Starting in one problem with a quadratic force and a pure sine wave as the initial position of the string, we indeed observe initially a gradual increase of energy in the higher modes as predicted (e.g., by Rayleigh in an infinitesimal analysis). Mode 2 starts increasing first, followed by mode 3, and so on. Later on, however, this gradual sharing of energy among successive modes ceases. ... It is only the first few modes which exchange energy among themselves and they do this in a rather regular fashion. Finally, at a later time mode 1 comes back to within one per cent of its initial value so that the system seems to be almost periodic. All our problems have at least this one feature in common. Instead of gradual increase of all the higher modes, the energy is exchanged, essentially, among only a certain few. It is, therefore, very hard to observe the rate of*

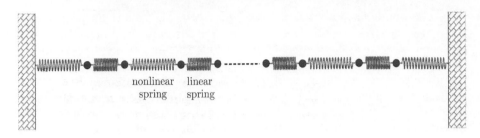

FIGURE 2.10: A Fermi-Pasta-Ulam system.

> *"thermalization" or mixing in our problem, and this was the initial*
> *purpose of the calculation.*

It is a shared feeling to think of the Fermi, Pasta and Ulam original investigations as the birth of experimental mathematics, that is the investigation of complex mathematical and physical problems (hard to be attacked by traditional methods) conducted with the aid of computer facilities.

The Fermi-Pasta-Ulam problem that will be the subject of our numerical investigations is a variant of the original one and was introduced by Galgani et al. in 1992. It models a physical system composed by $2m$ unit point masses disposed in series along a line and chained together by alternating weak nonlinear springs and stiff linear springs. In particular, we assume that the force exerted by the nonlinear springs are proportional to the cube of the displacement of the associated masses (cubic springs). The endpoints of the external springs are taken fixed (see Figure 2.10). We denote by q_1, q_2, ..., q_{2m} the displacements of the masses from their rest points and define the conjugate momenta as $p_i = \dot{q}_i$, $i = 1, \ldots, 2m$. The resulting problem is Hamiltonian and is defined by the energy function

$$
H(q, p) = \frac{1}{2} \sum_{i=1}^{m} (p_{2i-1}^2 + p_{2i}^2) + \underbrace{\frac{1}{4} \sum_{i=1}^{m} \omega_i^2 (q_{2i} - q_{2i-1})^2}_{\substack{\text{linear springs} \\ \text{potential}}} + \underbrace{\sum_{i=0}^{m} (q_{2i+1} - q_{2i})^4}_{\substack{\text{nonlinear springs} \\ \text{potential}}} \quad ,
$$

$$(2.51)$$

with $q_0 = q_{2m+1} = 0$ and where the coefficients ω_i, ruling the stiffness of the linear strings, may be large, thus yielding a *stiff oscillatory problem*.

As we will see in Chapter 4, our aim during the simulation of this system is to highlight the benefits of a Newton-like scheme versus a fixed point iteration to solve the nonlinear system at each step of the integration procedure. In fact, the stiff nature of system (2.51) obliges the fixed-point iteration, as well as any explicit numerical scheme suited for Hamiltonian problems, to use very small stepsizes.

2.8 Molecular dynamics

Molecular dynamics (MD) is a computer simulation of physical movements of atoms and molecules in the context of N-body simulation. It was originally conceived within theoretical physics, to simulate the properties of liquids, and then extended to other settings, such as computational chemistry and biology, where it is used in the simulation of biomolecules (e.g., for studying protein folding) and drug design.

Since complex molecular systems consist of a vast number of particles, it is usually impossible to study their properties analytically. MD simulation circumvents this problem by using numerical methods, based on the fact that, for systems which obey the ergodic hypothesis, the evolution of a single molecular dynamics simulation may be used to determine macroscopic thermodynamic properties of the system. However, this usually requires simulating the system over a relatively long period of time, so that MD is computationally challenging. As an example, in the context of drug design, the times of pharmaceutical interest are in the region of *msec* up to *min*, whereas simulation times are often in the region of *psec* up to *nsec* with *fsec* timesteps. This is partly due to the nature of the problem (in fact, the arising trajectories are often chaotic). Moreover, the numerical methods must be appropriately chosen, in order to correctly reproduce the macroscopic properties of the system.

In the practice, the atoms and molecules are allowed to interact for a period of time, giving a view of the motion of the atoms. In classical MD it is assumed that the motion of atoms and molecules can be described by Newtonian differential equations (as is done in classical mechanics), by replacing mechanical potentials by special molecular potentials. Consequently, the trajectories of atoms and molecules are determined by numerically solving Newton's equations of motion for a system of interacting particles, then leading to a Hamiltonian in the form (2.11)-(2.12), where forces between the particles and potential energy are defined by inter-atomic potentials or molecular mechanics force fields.

A simple mathematical model for such a potential is given by the *Lennard-Jones potential*, which approximates the interaction between a pair of neutral atoms or molecules placed at q_1 and q_2, respectively. It is given by

$$V_{LJ}(r) = 4\varepsilon \left[\left(\frac{\sigma}{r}\right)^{12} - \left(\frac{\sigma}{r}\right)^{6} \right], \tag{2.52}$$

where ε is the *potential well*, σ is the finite distance at which the inter-particle potential is zero, and $r := \|q_1 - q_2\|_2$ is the distance between the two particles. It is easy to check that V_{LJ} assumes its minimum value, $-\varepsilon$, at $r = r^* = 2^{1/6}\sigma$ (see Figure 2.11, for the case $\varepsilon = 10^{-3}$ and $\sigma = 1$).

In case of N particles, located at the positions $q_1, \ldots, q_N$, the above expression generalizes, leading to a Hamiltonian in the form (2.11)-(2.12), which

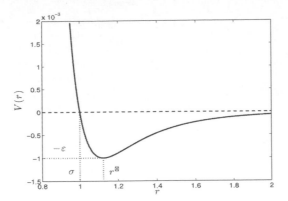

FIGURE 2.11: Lennard-Jones potential (2.52), $\varepsilon = 10^{-3}$, $\sigma = 1$.

reads

$$H(\boldsymbol{q}, \boldsymbol{p}) = \frac{1}{2} \sum_{i=1}^{N} \frac{\|p_i\|_2^2}{m_i} + 4 \sum_{1 \leqslant i < j \leqslant N} \varepsilon_{ij} \left[\left(\frac{\sigma_{ij}}{\|q_i - q_j\|_2} \right)^{12} - \left(\frac{\sigma_{ij}}{\|q_i - q_j\|_2} \right)^{6} \right],$$

(2.53)

with $p_i = m_i \dot{q}_i$ the momentum of the i-th particle, and ε_{ij} and σ_{ij} the potential well and the inter-particle 0-potential distance between particle i and j, respectively.

Bibliographical notes

The Hénon-Heiles problem has been introduced in [110]. Previous investigations on the problem are due to Contopoulos (see [78] and [79]).

For a thorough mathematical description of the circular restricted three-body problem see, for example, [4] and [168]. The first integral of the motion in the synodic reference system, i.e. (2.44), was introduced in 1836 by Jacobi [124] and bears his name. For the derivation of Hill's problem we have followed the approach discussed in [143].

The Fermi-Pasta-Ulam model was first formulated by Fermi, Pasta, and Ulam in [88] and consists of a chain of nonlinear springs sharing the same response law to applied forces. The model analyzed is a slight modification of the original one presented in [97] (see also [105]). For our discussion we have taken inspiration from [147], [15] and [145].

An introduction to molecular dynamics, with applications and methods, can be found in [133, 150, 154].

Chapter 3

Analysis of Hamiltonian Boundary Value Methods (HBVMs)

In this chapter we carry out a thorough convergence and stability analysis of Runge-Kutta line integral methods, the family of energy-conserving Runge-Kutta methods defined in Chapter 1, which represents the most relevant instance of line integral methods for Hamiltonian problems. In the present chapter these methods will be referred to as Hamiltonian Boundary Value Methods (HBVMs) but, as was observed in the remark on page 44, the two names are interchangeable since they identify the very same methods.

One of the simplest tools used for their derivation and analysis is provided by a local expansion of the right-hand side of the considered problem, along a suitable orthonormal basis. The truncation of the series is the first step towards the determination of the new formulae. In fact, their effective implementation on a computer cannot take aside the use of appropriate quadrature rules to handle the integrals appearing in their preliminary formulation.

This discretization process provides the final, ready for use, shape of HBVMs: they form a subclass of Runge-Kutta methods, characterized by a low-rank coefficient matrix, that can be thought of as a generalization of Gauss-Legendre collocation formulae. The related discussion, presented in the first two sections, forms the core of the chapter.

In Section 3.3 we discuss a number of relevant properties of HBVMs including a linear stability analysis and their connection with standard collocation methods.

Section 3.4 explores their close relationship with the integrators based on the least-square polynomial approximation to the continuous problem.

In Section 3.5, in addition to the analytical approach based on a local Fourier expansion (discussed in Section 3.1), we also show how HBVMs may be as well defined by exploiting a geometrical methodology related to the line integral in the phase space. Indeed, this latter tool has been the foundation for the genesis of the entire class of energy-preserving methods and, to maintain a historical perspective, we propose this analysis as a different, independent option to introduce the new formulae. This section also gathers a number of closely related and independent studies in the literature, such as the second order Averaged Vector Field method and its generalizations to higher order, as well as the time-finite element methods.

3.1   Derivation and analysis of the methods

Let us consider a problem in canonical Hamiltonian form,

$$\dot{y}(t) = J\nabla H(y(t)), \qquad t \geqslant 0, \qquad y(0) = y_0 \in \mathbb{R}^{2m}, \qquad (3.1)$$

where

$$J = \begin{pmatrix} & I_m \\ -I_m & \end{pmatrix}, \qquad (3.2)$$

with I_m denoting, as is usual, the identity matrix of dimension m, and H the Hamiltonian function defining the problem. As a shortcut, we shall also use the notation

$$f(y) := J\nabla H(y), \qquad (3.3)$$

both for brevity and because most of the arguments that we shall introduce hold true for a general differential equation of the form $\dot{y} = f(y)$.

Remark

For sake of simplicity, hereafter we shall assume that $f(y(t))$ can be expanded in Taylor series at $t = 0$.

Let us then introduce the following *uniform* partition, with *stepsize $h > 0$*:[1]

$$t_0 = 0, \qquad t_{n+1} = t_n + h, \qquad n = 0, 1, \ldots,$$

and define a procedure for getting corresponding approximations

$$y_n \approx y(t_n), \qquad n = 0, 1, \ldots,$$

to the solution of problem (3.1). Since we are going to define a *one-step* procedure, it will be enough to study its very first step of application, namely for (numerically) solving the *local problem*

$$\dot{y}(t) = f(y(t)), \qquad t \in [0, h], \qquad y(0) = y_0,$$

thus obtaining the approximation $y_1 \approx y(h)$. Once this will be done, the procedure is repeated iteratively, as explained by Algorithm 1.

---

[1] We shall assume the partition to be uniform, for sake of simplicity.

Algorithm 1

```
Given y₀
    For    n = 0, 1, ... :
            Compute    an approximation σ(t)
                       to the solution of the local problem
```
$$\begin{cases} \dot{\xi}(t) &= f(\xi(t)), \quad t \in [t_n, t_n + h], \\ \xi(t_n) &= y_n. \end{cases}$$
```
            Set        y_{n+1} := σ(t_n + h).
    End
```

Consequently, let us consider the very first local problem, which we transform on the interval $[0, 1]$:

$$\dot{y}(ch) = f(y(ch)), \qquad c \in [0, 1], \qquad y(0) = y_0. \tag{3.4}$$

We look for a suitable polynomial approximation $\sigma(ch)$ to the solution of (3.4), such that $\sigma(0) = y_0$. The numerical solution $y_1 \approx y(h)$ at time h is then defined as $y_1 := \sigma(h)$. For this purpose, we will exploit the family of Legendre polynomials, $\{P_j\}_{j \geqslant 0}$, scaled and shifted in order that

$$\deg P_j = j, \qquad \int_0^1 P_i(x) P_j(x) \mathrm{d}x = \delta_{ij}, \qquad \forall i, j = 0, 1, \ldots, \tag{3.5}$$

with δ_{ij} the Kronecker delta.[2] In other words, the polynomials are orthonormal, with respect to the L^2 scalar product

$$< g_1, g_2 > = \int_0^1 g_1(x) g_2(x) \mathrm{d}x, \qquad \forall g_1, g_2 \in L^2([0, 1]),$$

thus providing an orthonormal basis for such functions. One then obtains the following expansion along the orthonormal basis $\{P_j\}$,

$$f(y(ch)) = \sum_{j \geqslant 0} P_j(c) \gamma_j(y), \qquad c \in [0, 1], \tag{3.6}$$

with

$$\gamma_j(y) = \int_0^1 P_j(\tau) f(y(\tau h)) \mathrm{d}\tau, \qquad j \geqslant 0, \tag{3.7}$$

usually known as the Fourier-Legendre or Neumann expansion. Consequently, the solution of (3.4) can be formally written as

$$y(ch) = y_0 + h \sum_{j \geqslant 0} \int_0^c P_j(x) \mathrm{d}x \, \gamma_j(y), \qquad c \in [0, 1]. \tag{3.8}$$

---

[2] See Section A.1 for more details on Legendre polynomials.

In order to obtain a polynomial approximation $\sigma \in \Pi_s$ to $y(ch)$, we truncate the expansion (3.6) to a finite sum,

$$f(\sigma(ch)) \equiv \sum_{j \geq 0} P_j(c)\gamma_j(\sigma) \approx \sum_{j=0}^{s-1} P_j(c)\gamma_j(\sigma), \qquad c \in [0,1], \qquad (3.9)$$

with $\gamma_j(\sigma)$ formally defined by (3.7) with y replaced by σ:

$$\gamma_j(\sigma) = \int_0^1 P_j(\tau)f(\sigma(\tau h))\mathrm{d}\tau, \qquad j = 0,\ldots,s-1. \qquad (3.10)$$

Said differently, we approximate the local problem (3.4) by means of the following one,

$$\dot{\sigma}(ch) = \sum_{j=0}^{s-1} P_j(c)\gamma_j(\sigma), \qquad c \in [0,1], \qquad \sigma(0) = y_0, \qquad (3.11)$$

whose solution is given by (compare with (3.8))

$$\sigma(ch) = y_0 + h\sum_{j=0}^{s-1} \int_0^c P_j(x)\mathrm{d}x\,\gamma_j(\sigma), \qquad c \in [0,1]. \qquad (3.12)$$

Taking into account the orthogonal conditions in (3.5), and considering that $P_0(x) \equiv 1$, the new approximation $y_1 \approx y(h)$ is then given by

$$y_1 := \sigma(h) = y_0 + h\gamma_0(\sigma) \equiv y_0 + h\int_0^1 f(\sigma(\tau h))\mathrm{d}\tau. \qquad (3.13)$$

In order for the above approach to work, we first need to discuss the solvability of the nonlinear system (3.10). Using (3.12), we recast it as

$$\gamma_j = \int_0^1 P_j(\tau)f\left(y_0 + h\sum_{i=0}^{s-1} \int_0^\tau P_i(x)\mathrm{d}x\,\gamma_i\right)\mathrm{d}\tau, \qquad j = 0,\ldots,s-1, \qquad (3.14)$$

to make all the unknowns γ_j explicitly visible.[3]

Theorem 3.1. *Suppose $f : \mathbb{R}^n \to \mathbb{R}^n$ is Lipschitz continuous with constant $L > 0$. System (3.14) admits a unique solution provided that the stepsize h satisfies*

$$h < \frac{1}{L\max_{j=0,\ldots,s-1}\sum_{i=0}^{s-1}\int_0^1\left(|P_j(\tau)|\int_0^\tau |P_i(x)|\mathrm{d}x\right)\mathrm{d}\tau}. \qquad (3.15)$$

---

[3]Since the symbol σ denoting the polynomial approximation does not explicitly appear in (3.14), here and in the subsequent occurrences, we have replaced the notation $\gamma_j(\sigma)$ with the simpler notation γ_j.

Proof. We set γ and $\Psi(\gamma)$ the block vectors of $\mathbb{R}^{sn}$ whose j-th block entries are the left and right hand side of (3.14), respectively:

$$\gamma = \begin{pmatrix} \gamma_0 \\ \vdots \\ \gamma_{s-1} \end{pmatrix},$$ (3.16)

$$\Psi(\gamma) = \begin{pmatrix} \int_0^1 P_0(\tau) f(y_0 + h \sum_{i=0}^{s-1} \int_0^\tau P_i(x) \mathrm{d}x\, \gamma_i) \mathrm{d}\tau \\ \vdots \\ \int_0^1 P_{s-1}(\tau) f(y_0 + h \sum_{i=0}^{s-1} \int_0^\tau P_i(x) \mathrm{d}x\, \gamma_i) \mathrm{d}\tau \end{pmatrix}.$$

Consequently, system (3.14) is recast as $\gamma = \Psi(\gamma)$. Using the norm $\|\gamma\| = \max_j \|\gamma_j\|$ for vectors in $\mathbb{R}^{sn}$, a direct calculation shows that Ψ is Lipschitz continuous on $\mathbb{R}^{sn}$ with constant

$$K = hL \max_{j=0,\ldots,s-1} \sum_{i=0}^{s-1} \int_0^1 \left(|P_j(\tau)| \int_0^\tau |P_i(x)| \mathrm{d}x \right) \mathrm{d}\tau.$$ (3.17)

Thus, if h is small enough to satisfy (3.15), Ψ is a contraction and the Banach fixed-point theorem assures the existence of a unique solution of system (3.14) which can be found as the limit of the sequence $\gamma_{k+1} = \Psi(\gamma_k)$, $k = 0, 1, \ldots$. $\square$

Leaving aside for the moment the question of how to manage the integrals appearing in (3.14), the initial guess to start the iteration process could be, for example, $\gamma_0 = e_1 \otimes f(y_0)$ that is, the solution of (3.14) for $h = 0$.

Remark

As is the case with Runge-Kutta methods, if the function $f \in C^p$ in a neighborhood of y_0, the unique solvability of system (3.14) may be as well derived with the aid of the implicit function theorem, which also shows that the γ_j, viewed as functions of the stepsize h, inherit the same smoothness as the function f. Setting $G(h, \gamma) := \gamma - \Psi(\gamma)$, this is readily seen after noticing that $G(0, e_1 \otimes f(y_0)) = 0$ and $\partial G / \partial \gamma|_{h=0}$ is the identity matrix. In contrast, the contraction mapping principle provides us with a practical iterative scheme to approximate the solution at each step of the integration procedure.

Concerning energy conservation, the following result holds true.

Theorem 3.2. *With reference to (3.3)-(3.4) and (3.11)-(3.13), one has:*

$$H(y_1) = H(y_0).$$

Proof. By taking into account (3.7), one has:

$$
\begin{aligned}
H(y_1) - H(y_0) &= H(\sigma(h)) - H(\sigma(0)) = \int_0^h \nabla H(\sigma(t))^\mathsf{T} \dot\sigma(t)\mathrm{d}t \\
&= h\int_0^1 \nabla H(\sigma(ch))^\mathsf{T} \dot\sigma(ch)\mathrm{d}c = h\int_0^1 \nabla H(\sigma(ch))^\mathsf{T}\left[\sum_{j=0}^{s-1} P_j(c)\gamma_j(\sigma)\right]\mathrm{d}c \\
&= h\sum_{j=0}^{s-1}\left[\int_0^1 \underbrace{\nabla H(\sigma(ch))}_{J^\mathsf{T} f(\sigma(ch))} P_j(c)\mathrm{d}c\right]^\mathsf{T} \gamma_j(\sigma) \\
&= h\sum_{j=0}^{s-1}\underbrace{\left[\int_0^1 f(\sigma(ch))P_j(c)\mathrm{d}c\right]^\mathsf{T}}_{\gamma_j(\sigma)^\mathsf{T}} J\gamma_j(\sigma) = h\sum_{j=0}^{s=1} \gamma_j(\sigma)^\mathsf{T} J\gamma_j(\sigma) = 0,
\end{aligned}
$$

due to the fact that matrix J (see (3.2)) is skew-symmetric. $\qquad\square$

We now discuss the accuracy of the obtained approximation. For this purpose, we need some preliminary result.

Lemma 3.1. *Let $g : [0, h] \to V$, with V a vector space, admit a Taylor expansion at 0.[4] Then*

$$
\int_0^1 P_j(\tau)g(\tau h)\mathrm{d}\tau = O(h^j), \qquad j \geqslant 0.
$$

Proof. By hypothesis and the orthogonality conditions in (3.5), one has:

$$
\begin{aligned}
\int_0^1 P_j(\tau)g(\tau h)\mathrm{d}\tau &= \int_0^1 P_j(\tau)\sum_{k\geqslant 0}\frac{g^{(k)}(0)}{k!}(\tau h)^k\mathrm{d}\tau \\
&= \sum_{k\geqslant 0}\frac{g^{(k)}(0)}{k!}h^k\int_0^1 \tau^k P_j(\tau)\mathrm{d}\tau \\
&= \sum_{k\geqslant j}\frac{g^{(k)}(0)}{k!}h^k\int_0^1 \tau^k P_j(\tau)\mathrm{d}\tau = O(h^j). \quad\square
\end{aligned}
$$

Lemma 3.2. *With reference to the vectors defined at (3.7), one has:*

$$
\gamma_j(y) = O(h^j), \qquad \forall j \geqslant 0.
$$

Proof. Obvious, from Lemma 3.1, by setting $g = f \circ y$. $\qquad\square$

We also need the following standard perturbation result for initial value problems for ODEs, which we report without a proof.

---

[4]The cases of interest for our purposes are $V = \mathbb{R}^n$ and $V = \mathbb{R}^{m\times n}$.

Lemma 3.3. *Let $y(t; \tilde{t}, \tilde{y})$ be the solution (which we assume to exist) of the initial value problem*

$$\dot{y}(t) = f(y(t)), \qquad t \geq \tilde{t}, \qquad y(\tilde{t}) = \tilde{y}. \tag{3.18}$$

Then

$$\frac{\partial}{\partial \tilde{y}} y(t; \tilde{t}, \tilde{y}) = \Phi(t; \tilde{t}, \tilde{y}), \qquad \frac{\partial}{\partial \tilde{t}} y(t; \tilde{t}, \tilde{y}) = -\Phi(t; \tilde{t}, \tilde{y}) f(\tilde{y}),$$

with $\Phi(t; \tilde{t}, \tilde{y})$ the fundamental matrix, which is the solution of the variational problem

$$\dot{\Phi}(t; \tilde{t}, \tilde{y}) = f'(y(t; \tilde{t}, \tilde{y})) \Phi(t; \tilde{t}, \tilde{y}), \qquad t \geq \tilde{t}, \qquad \Phi(\tilde{t}; \tilde{t}, \tilde{y}) = I.$$

Theorem 3.3. *With reference to (3.4) and (3.11)–(3.13), one has:*

$$y_1 - y(h) = O(h^{2s+1}).$$

Proof. By setting $y(t; \tilde{t}, \tilde{y})$ the solution of problem (3.18), one has:

$$
\begin{aligned}
y_1 - y(h) &= \sigma(h) - y(h) \\
&= y(h; h, \sigma(h)) - y(h; 0, \sigma(0)) = \int_0^h \frac{\mathrm{d}}{\mathrm{d}t} y(h; t, \sigma(t)) \, \mathrm{d}t \\
&= \int_0^h \left[\frac{\partial}{\partial \tilde{t}} y(h; \tilde{t}, \sigma(t)) \Big|_{\tilde{t}=t} + \frac{\partial}{\partial \tilde{y}} y(h; t, \tilde{y}) \Big|_{\tilde{y}=\sigma(t)} \dot{\sigma}(t) \right] \mathrm{d}t \\
&= \int_0^h \underbrace{\left[-\Phi(h; t, \sigma(t)) f(\sigma(t)) + \Phi(h; t, \sigma(t)) \dot{\sigma}(t) \right]}_{\text{by Lemma 3.3}} \mathrm{d}t \\
&= h \int_0^1 \left[-\Phi(h; ch, \sigma(ch)) f(\sigma(ch)) + \Phi(h; ch, \sigma(ch)) \dot{\sigma}(ch) \right] \mathrm{d}c \\
&= -h \int_0^1 \Phi(h; ch, \sigma(ch)) \underbrace{\left[\sum_{j \geq 0} P_j(c) \gamma_j(\sigma) - \sum_{j=0}^{s-1} P_j(c) \gamma_j(\sigma) \right]}_{\text{by (3.9) and (3.11)}} \mathrm{d}c \\
&= -h \int_0^1 \Phi(h; ch, \sigma(ch)) \sum_{j \geq s} P_j(c) \gamma_j(\sigma) \mathrm{d}c \\
&= -h \sum_{j \geq s} \underbrace{\left[\int_0^1 P_j(c) \Phi(h; ch, \sigma(ch)) \mathrm{d}c \right]}_{O(h^j) \text{ by Lemma 3.1}} \overbrace{\gamma_j(\sigma)}^{O(h^j) \text{ by Lemma 3.2}} = O(h^{2s+1}).
\end{aligned}
$$

□

The following results are direct consequences of the previous theorem.

Corollary 3.1. *Integrating on a finite time interval, the sequence* $\{y_n\}$ *provided by Algorithm 1 on page 83 has order of convergence 2s towards the analytical solution, as* $h \to 0$.

Corollary 3.2. *Set*

$$G(ch) \equiv (g_{ij}(ch)) := \Phi(h; ch, \sigma(ch)) \quad and \quad w(ch) \equiv (w_i(ch)) := f(\sigma(ch)).$$

Then, the local truncation error may be cast as

$$
y_1 - y(h) = -h^{2s+1} \left(\frac{s!}{(2s)!\sqrt{2s+1}} \right)^2 \left(g_{ij}^{(s)}(\xi_{ij}h) \right) \left(w_i^{(s)}(\eta_i h) \right)
$$
$$
+ O(h^{2s+2}), \tag{3.19}
$$

where ξ_{ij} *and* η_i *are suitable points in the interval* $(0, 1)$. *The first term at the right-hand side of (3.19) represents the leading term in the expansion.*

Proof. From the last line of the proof of Theorem 3.3, one obtains:

$$
y_1 - y(h) = -h \int_0^1 P_s(c)\Phi(h; ch, \sigma(ch)) \mathrm{d}c \, \gamma_s(\sigma) + O(h^{2s+2})
$$
$$
= -h \int_0^1 P_s(c)\Phi(h; ch, \sigma(ch)) \mathrm{d}c \int_0^1 P_s(c) f(\sigma(ch)) \mathrm{d}c + O(h^{2s+2})
$$
$$
\equiv -h \int_0^1 P_s(c) G(ch) \mathrm{d}c \int_0^1 P_s(c) w(ch) \mathrm{d}c + O(h^{2s+2}). \tag{3.20}
$$

To estimate the two integrals, we consider the Gaussian quadrature formula of order $2s$. Its nodes $c_1, \ldots, c_s$ are just the roots of $P_s(c)$, thus the quadrature formulae vanish and the integrals match the error in the approximation. Consequently, by considering that $P_s^{(k)}(c) \equiv 0$ for $k > s$, we get:

$$
\int_0^1 P_s(c) g_{ij}(ch) \mathrm{d}c = \frac{(s!)^4}{(2s+1)\left[(2s)!\right]^3} \frac{d^{2s}}{dc^{2s}} \left[P_s(c) g_{ij}(ch) \right]_{c=\xi_{ij}}
$$
$$
= \frac{(s!)^4}{(2s+1)\left[(2s)!\right]^3} \sum_{k=0}^{2s} \binom{2s}{k} \left[P_s^{(k)}(c) \frac{d^{2s-k}}{dc^{2s-k}} g_{ij}(ch) \right]_{c=\xi_{ij}}
$$
$$
= \frac{(s!)^4}{(2s+1)\left[(2s)!\right]^3} \sum_{k=0}^{s} \binom{2s}{k} \left[P_s^{(k)}(c) g_{ij}^{(2s-k)}(ch) \right]_{c=\xi_{ij}} h^{2s-k}
$$
$$
= h^s \frac{(s!)^4}{(2s+1)\left[(2s)!\right]^3} \binom{2s}{s} \left[P_s^{(s)}(c) g_{ij}^{(s)}(ch) \right]_{c=\xi_{ij}} + O(h^{s+1}),
$$

where ξ_{ij} is a suitable point in $(0, 1)$. Considering that $P_s^{(s)}(c) = s!\sqrt{2s+1}\binom{2s}{s}$ (see property **P5** in Section A.1), it follows that

$$
\int_0^1 P_s(c) g_{ij}(ch) \mathrm{d}c = h^s \frac{s!}{(2s)!\sqrt{2s+1}} g_{ij}^{(s)}(\xi_{ij}h) + O(h^{s+1}).
$$

Analogously, for the second integral we obtain, for suitable $\eta_i \in (0,1)$,

$$\int_0^1 P_s(c)w_i(ch)dc = h^s \frac{s!}{(2s)!\sqrt{2s+1}} w_i^{(s)}(\eta_i h) + O(h^{s+1}).$$

Substituting the values of these two integrals back into (3.20) yields the final result.   □

3.1.1   Discretization

Formulae (3.13)-(3.14), though characterizing the polynomial approximation $\sigma(ch)$, and consequently the numerical solution $y_1 = \sigma(h) \approx y(h)$, do not provide a numerical method yet. To make them ready for use, one has to approximate the integrals involving the function f: this can be done by using a polynomial interpolation formula based at the k abscissae

$$0 \leqslant c_1 < \cdots < c_k \leqslant 1, \tag{3.21}$$

with corresponding weights $b_1, \ldots, b_k$. In so doing, the polynomial approximation generally changes, due to possible truncation errors in the quadrature: let us denote $u(ch) \in \Pi_s$ the new approximation. The numerical method, generated by approximating the integrals in (3.13)-(3.14) with the quadrature (c_i, b_i), then reads:

$$u(ch) \;=\; y_0 + h \sum_{j=0}^{s-1} \int_0^c P_j(x)dx\, \hat{\gamma}_j, \qquad c \in [0,1], \tag{3.22}$$

$$\hat{\gamma}_j \;=\; \sum_{i=1}^k b_i P_j(c_i) f(u(c_i h)), \qquad j = 0, \ldots, s-1, \tag{3.23}$$

$$y_1 \;:=\; u(h) \;\equiv\; y_0 + h \sum_{i=1}^k b_i f(u(c_i h)). \tag{3.24}$$

We now study how the introduction of the quadrature formulae, in place of the integrals, affects the convergence and energy conservation results presented in Section 3.1. Clearly, they will depend on the accuracy of the quadrature formula, whose order we denote by q, which means that it is exact for polynomial integrands up to degree $q - 1$.

Remark

By placing the abscissae (3.21) at the Gauss-Legendre nodes in $[0, 1]$, i.e., at the zeros of P_k,

$$P_k(c_i) = 0, \qquad i = 1, \ldots, k, \tag{3.25}$$

the order of the quadrature is maximum: $q = 2k$. According to property **P2** in Section A.1, the corresponding weights are given by:

$$b_i = \frac{4(2k-1)c_i(1-c_i)}{[kP_{k-1}(c_i)]^2}, \qquad i = 1, \ldots, k.$$

Hereafter, we shall always assume this choice.

Lemma 3.4. *Assume that the interpolation quadrature formula (c_i, b_i) is based at the abscissae (3.25). Consider the polynomial $u(ch)$ defined at (3.22) and the vectors*

$$\gamma_j(u) := \int_0^1 P_j(c)f(u(ch))dc, \quad j = 0, \ldots, s-1.$$

Then, for all $k \geqslant s$ and $j = 0, \ldots, s-1$:

$$\begin{aligned} \Delta_j(h) &:= \gamma_j(u) - \hat{\gamma}_j \\[4pt] &= \begin{cases} 0, & \text{if } f \in \Pi_{\nu-1}, \text{ with } \nu \leqslant 2k/s, \\[4pt] O(h^{2k-j}), & \text{otherwise.} \end{cases} \end{aligned} \tag{3.26}$$

Proof. The statement follows from the fact that, if $f \in \Pi_{\nu-1}$, then

$$\Delta_j(h) \equiv \int_0^1 P_j(c)f(u(ch))dc - \sum_{i=1}^k b_i P_j(c_i)f(u(c_ih)) = 0, \tag{3.27}$$

since the integrand has at most degree (recall that $u \in \Pi_s$)

$$s - 1 + (\nu - 1)s = \nu s - 1 \leqslant 2k - 1,$$

so that the quadrature is exact. Conversely, one has, by setting $g = f \circ u$,

$$\begin{aligned} \Delta_j(h) &\propto \frac{\mathrm{d}^{2k}}{\mathrm{d}c^{2k}}[P_j(c)f(u(ch))] = \frac{\mathrm{d}^{2k}}{\mathrm{d}c^{2k}}[P_j(c)g(ch)] \\[6pt] &= \sum_{i=0}^{2k}\binom{2k}{i}P_j^{(i)}(c)g^{(2k-i)}(ch)h^{2k-i} \\[6pt] &= \sum_{i=0}^{j}\binom{2k}{i}P_j^{(i)}(c)g^{(2k-i)}(ch)h^{2k-i} = O(h^{2k-j}), \end{aligned}$$

since the derivatives of order greater than j of $P_j(c)$ vanish. $\qquad \square$

This result allows us to discuss how the discretization process influences the energy-conservation property of the methods. As one can easily argue, increasing the order of the quadrature far beyond $2s$ improves the ability of the full discretized formula in preserving the energy.

Theorem 3.4. *With reference to (3.22)–(3.25), one has that, for all $k \geqslant s$:*[5]

$$H(y_1) - H(y_0) = \begin{cases} 0, & \text{if } H \in \Pi_\nu, \quad \text{with} \quad \nu \leqslant 2k/s, \\ O(h^{2k+1}), & \text{otherwise.} \end{cases}$$

Proof. The proof, based on the use of a line integral, is similar to that of Theorem 3.2:

$$H(y_1) - H(y_0) = H(u(h)) - H(u(0)) = \int_0^h \nabla H(u(t))^\top \dot{u}(t) dt$$

$$= h \int_0^1 \nabla H(u(ch))^\top \dot{u}(ch) dc = h \int_0^1 \nabla H(u(ch))^\top \left[\sum_{j=0}^{s-1} P_j(c)\hat{\gamma}_j \right] dc$$

$$= h \sum_{j=0}^{s-1} \left[\int_0^1 J \nabla H(u(ch)) P_j(c) dc \right]^\top J\hat{\gamma}_j$$

$$= h \sum_{j=0}^{s-1} \gamma_j(u)^\top J \left[\gamma_j(u) - \Delta_j(h) \right] = -h \sum_{j=0}^{s-1} \gamma_j(u)^\top J \Delta_j(h).$$

Our claim then follows by observing that $\gamma_j(u) = O(h^j)$ and, by virtue of Lemma 3.4,

$$\Delta_j(h) = \begin{cases} 0, & \text{if } H \in \Pi_\nu, \quad \text{with} \quad \nu \leqslant 2k/s, \\ O(h^{2k-j}), & \text{otherwise.} \end{cases}$$

$\square$

Remark

Theorem 3.4 allows us to conclude that, by choosing k large enough, energy conservation can always be gained:

- *exactly*, for all polynomial Hamiltonians;

- *practically*, for all suitably regular Hamiltonians.

In fact, in the latter case it will suffice to bring the quadrature error within round-off.

---

[5]For a general quadrature formula of order q, one would obtain that $H(y_1) - H(y_0) = 0$, if $H \in \Pi_\nu$, with $\nu \leqslant q/s$, or $H(y_1) - H(y_0) = O(h^{q+1})$, otherwise.

Let us turn to the question about the accuracy of the new approximation (3.24).

Theorem 3.5. *With reference to (3.22)–(3.25), one has that, for all $k \geqslant s$:*

$$y_1 - y(h) = O(h^{2s+1}).$$

Namely, the method has order $2s$.[6]

Proof. We proceed along the lines of the proof of Theorem 3.3, by simply replacing $\sigma(ch)$ with $u(ch)$. By virtue of (3.26) we get:

$$
\begin{aligned}
y_1 - y(h) \;&=\; u(h) - y(h) \\[2mm]
&=\; y(h; h, u(h)) - y(h; 0, u(0)) \;=\; \int_0^h \frac{\mathrm{d}}{\mathrm{d}t} y(h; t, u(t))\, \mathrm{d}t \\[2mm]
&=\; \int_0^h \left[\frac{\partial}{\partial \tilde{t}}\, y(h; \tilde{t}, u(t))\big|_{\tilde{t}=t} + \frac{\partial}{\partial \tilde{y}}\, y(h; t, \tilde{y})\big|_{\tilde{y}=u(t)}\, \dot{u}(t) \right] \mathrm{d}t \\[2mm]
&=\; \int_0^h \left[-\Phi(h; t, u(t)) f(u(t)) + \Phi(h; t, u(t)) \dot{u}(t) \right] \mathrm{d}t \\[2mm]
&=\; h \int_0^1 \left[-\Phi(h; ch, u(ch)) f(u(ch)) + \Phi(h; ch, u(ch)) \dot{u}(ch) \right] \mathrm{d}c \\[2mm]
&=\; -h \int_0^1 \Phi(h; ch, u(ch)) \left[\sum_{j \geqslant 0} P_j(c)\gamma_j(u) - \sum_{j=0}^{s-1} P_j(c)\hat{\gamma}_j \right] \mathrm{d}c \\[2mm]
&=\; h \int_0^1 \Phi(h; ch, u(ch)) \left[-\sum_{j=0}^{s-1} P_j(c)\Delta_j(h) - \sum_{j \geqslant s} P_j(c)\gamma_j(u)\mathrm{d}c \right] \\[2mm]
&=\; -h \sum_{j=0}^{s-1} \left[\int_0^1 P_j(c)\Phi(h; ch, u(ch))\mathrm{d}c \right] \underbrace{\overbrace{\Delta_j(h)}^{O(h^{2k-j})}}_{O(h^j)} \\[2mm]
&\quad\; -h \sum_{j \geqslant s} \left[\int_0^1 P_j(c)\Phi(h; ch, u(ch))\mathrm{d}c \right] \gamma_j(u) \qquad\qquad (3.28) \\[2mm]
&=\; O(h^{2k+1}) + O(h^{2s+1}) \;=\; O(h^{2s+1}).
\end{aligned}
$$

$\square$

We then conclude that the order of the new approximation is the same as that of the original formulae (3.13)-(3.14), i.e., $2s$, provided that the order of the quadrature be $q \geqslant 2s$.

---

[6] For a general quadrature formula of order q, one would obtain that $y_1 - y(h) = O(h^{p+1})$, with $p = \min\{q, 2s\}$.

We observe that the leading term in the local truncation error of the method may be conveniently deduced from (3.28). In particular, two cases may arise:

(1) Choosing $k > s$ makes the first sum in (3.28) negligible with respect to the second as $h \to 0$. Since this latter sum is formally the same as the one we encountered at the end of the proof of Theorem 3.3, we conclude that the leading term is again given by formula (3.19) in Corollary 3.2, with $G(ch)$ and $w(ch)$ defined in terms of the polynomial $u(ch)$ instead of $\sigma(ch)$.

(2) The choice $k = s$ means that we are using the Gauss collocation method of order $2s$.[7] In such a case, all the terms in the first sum of (3.28) bring a $O(h^{2s+1})$ contribution and cannot be neglected anymore. Exploiting the same flow of computations as in the proof of Corollary 3.2, it is easy to deduce the following estimations, for $j = 0, \ldots, s-1$:

$$\int_0^1 P_j(c)\Phi(h;ch,u(ch))\mathrm{d}c = h^j \frac{j!}{(2j)!\sqrt{2j+1}}\widehat{G}_j + O(h^{j+1})$$

obtained by approximating the integral by means of the Gaussian quadrature rule of order $2j$. Matrices $\widehat{G}_j$ are defined in terms of matrix

$$G(ch) \equiv (g_{i\ell}(ch)) := \Phi(h;ch,u(ch))$$

as $\widehat{G}_j := \left(g_{i\ell}^{(2j)}(\xi_{i\ell}^{(j)}h)\right)$, for suitable $\xi_{i\ell}^{(j)} \in (0,1)$. On the other hand, setting

$$w(ch) \equiv (w_i(ch)) := f(u(ch))$$

and considering (3.26)-(3.27) with $k = s$, for the i-th component of vector $\Delta_j(h)$ we get, for suitable $\eta_i^{(j)} \in (0,1)$,

$$
\begin{aligned}
(\Delta_j(h))_i &= \frac{(s!)^4}{(2s+1)[(2s)!]^3}\frac{\mathrm{d}^{2s}}{\mathrm{d}c^{2s}}\Big[P_j(c)w_i(ch)\Big]_{c=\eta_i^{(j)}} \\
&= \frac{(s!)^4}{(2s+1)[(2s)!]^3}\sum_{\ell=0}^{2s}\binom{2s}{\ell}\Big[P_j^{(\ell)}(c)\frac{\mathrm{d}^{2s-\ell}}{\mathrm{d}c^{2s-\ell}}w_i(ch)\Big]_{c=\eta_i^{(j)}} \\
&= \frac{(s!)^4}{(2s+1)[(2s)!]^3}\sum_{\ell=0}^{j}\binom{2s}{\ell}\Big[P_j^{(\ell)}(c)w_i^{(2s-\ell)}(ch)\Big]_{c=\eta_i^{(j)}}h^{2s-\ell} \\
&= h^{2s-j}\frac{(s!)^4}{(2s+1)[(2s)!]^3}\binom{2s}{j}\binom{2j}{j}j!\sqrt{2j+1}\,w_i^{(2s-j)}(\eta_i^{(j)}h) \\
&\quad + O(h^{2s-j+1}).
\end{aligned}
$$

---

[7]See Theorem 3.7 in the next section.

In conclusion, for the first term of (3.28) we have

$$h \sum_{j=0}^{s-1} \left[\int_0^1 P_j(c)\Phi(h; ch, u(ch))\mathrm{d}c \right] \Delta_j(h)$$

$$= h^{2s+1} \frac{(s!)^4}{(2s+1)[(2s)!]^3} \sum_{j=0}^{s-1} \binom{2s}{j} \widehat{G}_j \widehat{w}_j, \qquad (3.29)$$

where the vectors $\widehat{w}_j$ are defined as $\widehat{w}_j := \left(w_i^{(2s-j)}(\eta_i^{(j)}h) \right)$. Of course, the second term in (3.28) can be again estimated by formula (3.19).

From the analysis of the above two cases, one deduces that using $k > s$ gives usually a more favorable truncation error, but that is not likely the rule since, in principle, the presence of (3.29) could reduce the size of the resulting asymptotic error constant.

Definition 3.1. *We name* Hamiltonian Boundary Value Method with k stages and degree s, *in short HBVM(k, s), the methods defined by (3.22)–(3.25).*

In view of the result of Theorem 3.5, hereafter we shall always assume $k \geqslant s$. Moreover, we observe that, for fixed s, and assuming as usual that H is suitably regular, the polynomial u defined by a HBVM(k, s) method and the polynomial σ defined by (3.10) and (3.12) are obviulsy related as follows:

$$u(ch) \to \sigma(ch) \quad \text{uniformly}, \qquad \text{as} \qquad k \to \infty.$$

In other words, we could associate the polynomial σ to the method

$$\mathrm{HBVM}(\infty, s) := \lim_{k \to \infty} \mathrm{HBVM}(k, s). \qquad (3.30)$$

3.2   Runge-Kutta formulation

In this section, we show that a HBVM(k, s) method, with $k \geqslant s$, is nothing but a k-stage implicit Runge-Kutta method which generalizes the basic s-stage Gauss-Legendre collocation formula. We start by observing that, to advance the solution, the method defined at (3.22)–(3.24) only takes information from the values that the polynomial $u(ch)$ assumes at the abscissae $c_1, \ldots, c_k$, the explicit computation of the polynomial itself being unnecessary. Setting

$$Y_i := u(c_i h), \qquad i = 1, \ldots, k,$$

from (3.22)–(3.24) we obtain:

$$
\begin{aligned}
Y_i &= y_0 + h \sum_{j=0}^{s-1} \int_0^{c_i} P_j(x)\mathrm{d}x \sum_{\ell=1}^{k} b_\ell P_j(c_\ell) f(Y_\ell) \\
&= y_0 + h \sum_{j=1}^{k} b_j \left[\sum_{\ell=0}^{s-1} P_\ell(c_j) \int_0^{c_i} P_\ell(x)\mathrm{d}x \right] f(Y_j), \quad i = 1, \ldots, k, \\
y_1 &= y_0 + h \sum_{i=1}^{k} b_i f(Y_i).
\end{aligned}
\tag{3.31}
$$

In other words, we are speaking of the k-stage Runge-Kutta method with the following Butcher tableau:

$$
\begin{array}{c|c}
\boldsymbol{c} & A \\
\hline
& \boldsymbol{b}^\top
\end{array},
\tag{3.32}
$$

$$
\boldsymbol{b} = \begin{pmatrix} b_1 \\ \vdots \\ b_k \end{pmatrix}, \quad
\boldsymbol{c} = \begin{pmatrix} c_1 \\ \vdots \\ c_k \end{pmatrix}, \quad
A = \left(b_j \sum_{\ell=0}^{s-1} P_\ell(c_j) \int_0^{c_i} P_\ell(x)\mathrm{d}x \right)_{i,j=1,\ldots,k}.
$$

Let us now derive a more compact expression for the Butcher array $A \in \mathbb{R}^{k \times k}$. For this purpose, we introduce the following matrices:

$$
\mathcal{I}_s := \begin{pmatrix}
\int_0^{c_1} P_0(x)\mathrm{d}x & \cdots & \int_0^{c_1} P_{s-1}(x)\mathrm{d}x \\
\vdots & & \vdots \\
\int_0^{c_k} P_0(x)\mathrm{d}x & \cdots & \int_0^{c_k} P_{s-1}(x)\mathrm{d}x
\end{pmatrix}
$$

$$
\equiv \left(\int_0^{c_i} P_{j-1}(x)\mathrm{d}x \right) \in \mathbb{R}^{k \times s},
\tag{3.33}
$$

$$
\mathcal{P}_r := \begin{pmatrix}
P_0(c_1) & \cdots & P_{r-1}(c_1) \\
\vdots & & \vdots \\
P_0(c_k) & \cdots & P_{r-1}(c_k)
\end{pmatrix}
$$

$$
\equiv \left(P_{j-1}(c_i) \right) \in \mathbb{R}^{k \times r}, \quad r = s, s+1,
\tag{3.34}
$$

and

$$
\Omega := \begin{pmatrix} b_1 & & \\ & \ddots & \\ & & b_k \end{pmatrix} \in \mathbb{R}^{k \times k}.
\tag{3.35}
$$

The following result then holds true.

Theorem 3.6. *The matrix A of the Butcher tableau (3.32) can be written as* $A = \mathcal{I}_s \mathcal{P}_s^\top \Omega$.

Proof. Let us denote $e_\ell \in \mathbb{R}^k$ the ℓ-th vector of the canonical basis. We have:

$$e_i^\top (I_s P_s^\top \Omega) e_j = (e_i^\top I_s)(P_s^\top e_j) b_j$$

$$= b_j \left(\int_0^{c_i} P_0(x)\mathrm{d}x, \quad \dots, \quad \int_0^{c_i} P_{s-1}(x)\mathrm{d}x \right) \begin{pmatrix} P_0(c_j) \\ \vdots \\ P_{s-1}(c_j) \end{pmatrix}$$

$$= b_j \sum_{\ell=0}^{s-1} P_\ell(c_j) \int_0^{c_i} P_\ell(x)\mathrm{d}x \equiv (A)_{ij}. \qquad \square$$

The following preliminary results will be exploited later to state further properties of the Butcher array A.

Lemma 3.5. *With reference to (3.33)-(3.34), one has:*

$$I_s = P_{s+1} \hat{X}_s \equiv P_{s+1} \left(\frac{X_s}{(0 \ \dots \ 0 \ \xi_s)} \right), \tag{3.36}$$

with

$$X_s = \begin{pmatrix} \frac{1}{2} & -\xi_1 & & \\ \xi_1 & 0 & \ddots & \\ & \ddots & \ddots & -\xi_{s-1} \\ & & \xi_{s-1} & 0 \end{pmatrix} \in \mathbb{R}^{s \times s}, \quad \xi_i = \frac{1}{2\sqrt{4i^2 - 1}}. \tag{3.37}$$

Proof. The statement follows from property **P9** in Section A.1, which relates the integrals of the Legendre polynomials with the polynomials themselves. $\square$

Lemma 3.6. *The following properties of matrices (3.33)–(3.35) and (3.37) hold true for $k \geqslant s$:*

- $P_s^\top \Omega P_s = I_s$;
- *when $k = s$, $P_s^\top \Omega = P_s^{-1}$;*
- $P_s^\top \Omega I_s = X_s$.

Proof. The proof of the first property follows from the fact that, since $k \geqslant s$, the quadrature (c_i, b_i) is exact for polynomial integrands whose degree does not exceed $2s - 1$. Consequently, by setting $e_\ell \in \mathbb{R}^s$ the ℓ-th vector of the canonical basis, one has:

$$e_i^\top (P_s^\top \Omega P_s) e_j = (P_s e_i)^\top \Omega (P_s e_j) = \sum_{\ell=1}^k b_\ell P_{i-1}(c_\ell) P_{j-1}(c_\ell)$$

$$= \int_0^1 P_{i-1}(x) P_{j-1}(x) \mathrm{d}x = \delta_{ij}, \quad i, j = 1, \dots, s.$$

The second property easily follows from the first one, since when $k = s$ the matrix $\mathcal{P}_s$ is square. Concerning the third point, we first observe that

$$\mathcal{P}_s^\top \Omega \mathcal{P}_{s+1} = (\; I_s \quad \mathbf{0} \;).$$

In fact,

$$\mathcal{P}_s^\top \Omega \mathcal{P}_{s+1} = \mathcal{P}_s^\top \Omega (\; \mathcal{P}_s \quad \mathbf{p}_s \;), \quad \text{with } \mathbf{p}_s = (\; P_s(c_1) \quad \cdots \quad P_s(c_k) \;)^\top,$$

and the same arguments used in the first point yield, for $i = 1, \ldots, s$:

$$e_i^\top (\mathcal{P}_s^\top \Omega \mathbf{p}_s) = (\mathcal{P}_s e_i)^\top \Omega \mathbf{p}_s$$

$$= \sum_{\ell=1}^k b_\ell P_{i-1}(c_\ell) P_s(c_\ell) = \int_0^1 P_{i-1}(x) P_s(x) \mathrm{d}x = 0.$$

Thus, from Lemma 3.5 and (3.36)-(3.37), we finally obtain:

$$\mathcal{P}_s^\top \Omega \mathcal{I}_s = \mathcal{P}_s^\top \Omega \mathcal{P}_{s+1} \hat{X}_s = (\; I_s \quad \mathbf{0} \;) \hat{X}_s = X_s. \quad \square$$

Exploiting (3.25), Theorem 3.6, and Lemmas 3.5 and 3.6, the following result is readily established.

Theorem 3.7. *When $k = s$, the Butcher tableau of the HBVM(s, s) method is given by:*

$$\frac{c \;\;|\; \mathcal{P}_s X_s \mathcal{P}_s^{-1}}{\quad\; \mathbf{b}^\top}, \qquad \mathbf{b} = \begin{pmatrix} b_1 \\ \vdots \\ b_s \end{pmatrix}, \qquad \mathbf{c} = \begin{pmatrix} c_1 \\ \vdots \\ c_s \end{pmatrix}. \tag{3.38}$$

When $k > s$, the Butcher tableau of the HBVM(k, s) method reads:

$$\frac{c \;\;|\; \mathcal{P}_{s+1} \hat{X}_s \mathcal{P}_s^\top \Omega}{\quad\; \mathbf{b}^\top}, \qquad \mathbf{b} = \begin{pmatrix} b_1 \\ \vdots \\ b_k \end{pmatrix}, \qquad \mathbf{c} = \begin{pmatrix} c_1 \\ \vdots \\ c_k \end{pmatrix}. \tag{3.39}$$

Remark

From (3.38), we recognize the s-stage Gauss-Legendre collocation method written with the aid of the so called W-transformation studied in Section 1.4.3: namely the transformation provided by matrix $\mathcal{P}_s$. Thus, (3.39) provides a generalization of the W-transformation to the case of rectangular matrices.

Consequently, for $k \geqslant s$, HBVM(k, s) methods can be regarded as a generalization of the basic s-stage Gauss-Legendre collocation method.

3.3   Properties of HBVMs

It is interesting to observe that, by virtue of Theorem 3.5, for s fixed, all HBVM(k, s) methods share the same order $2s$, for all $k \geqslant s$, the only difference stemming from the result of Theorem 3.4, concerning the conservation of the Hamiltonian. Moreover, with Gauss collocation methods HBVMs also share a number of additional relevant properties, which we study in this section.

3.3.1   Isospectrality

To begin with, we state the following result, which can be easily proved by induction.

Lemma 3.7. *The determinant of matrix X_s in (3.37) is given by:*

$$\det(X_s) = \begin{cases} \prod_{i=1}^{s/2} \xi_{2i-1}^2, & \text{if } s \text{ is even;} \\ \frac{1}{2} \prod_{i=1}^{(s-1)/2} \xi_{2i}^2, & \text{if } s \text{ is odd.} \end{cases}$$

Consequently, matrix X_s is nonsingular for all $s \geqslant 1$.

With the aid of the above result we can now state the following spectral property of the coefficient matrix associated with a HBVM(k, s) method.

Theorem 3.8. *For all $k \geqslant s$, the Butcher array of a HBVM(k, s) method:*

- *has rank s;*

- *has the nonzero eigenvalues coinciding with those of X_s (plus a $(k - s)$-fold zero eigenvalue).*

Proof. The result holds trivially when $k = s$. For $k > s$, we observe that matrix $\mathcal{P}_s$ has full rank, i.e., s.[8] Moreover, the weights of the quadrature are positive, so Ω is nonsingular. Consequently,

$$\begin{aligned} s \geqslant \operatorname{rank}(A) &= \operatorname{rank}(\mathcal{P}_{s+1}\hat{X}_s\mathcal{P}_s^\top\Omega) \\ &\geqslant \operatorname{rank}(\underbrace{\mathcal{P}_s^\top\Omega\mathcal{P}_{s+1}\hat{X}_s}_{=X_s}\overbrace{\mathcal{P}_s^\top\Omega\mathcal{P}_s}^{=I_s}) = \operatorname{rank}(X_s) = s. \end{aligned}$$

This proves the first point. The second property follows by observing that

$$\mathcal{P}_s^\top\Omega A = \mathcal{P}_s^\top\Omega(\mathcal{P}_{s+1}\hat{X}_s\mathcal{P}_s^\top\Omega) = X_s\mathcal{P}_s^\top\Omega \quad \Longleftrightarrow \quad A^\top\Omega\mathcal{P}_s = \Omega\mathcal{P}_sX_s^\top.$$

This means that the columns of $\Omega\mathcal{P}_s$ span an s-dimensional (right) invariant subspace of $A^\top$. Therefore, the eigenvalues of X_s will coincide with the nonzero eigenvalues of A, the remaining ones being zero, since A has rank s. $\square$

---

[8]It is, indeed, a Gram-type matrix, defined by s linearly independent polynomials.

Since, for fixed s, the coefficient matrices

$$A = \mathcal{P}_{s+1}\hat{X}_s \mathcal{P}_s^\top \Omega \in \mathbb{R}^{k \times k}, \quad k = s, s+1, \ldots,$$

of HBVM(k, s) methods share the same nonzero eigenvalues (which, in turn, coincide with those of the coefficient matrix of the s-stage Gauss-Legendre collocation method), we say that such methods possess the *isospectral property*.

3.3.2 Equivalence property for polynomial problems

This property is a direct consequence of the discretization process described in Section 3.1.1 and pertaining to the approximation of the integrals involved in the definition of the polynomial curve $\sigma(ch) \in \Pi_s$, namely

$$\gamma_j(\sigma) = \int_0^1 P_j(\tau) f(\sigma(\tau h)) \mathrm{d}\tau, \quad j = 0, \ldots, s - 1,$$

by means of general interpolation quadrature formulae, which means that the distribution of the nodes and their number may be arbitrary.

In the event that the function f is a polynomial of degree, say $\nu - 1$, a quadrature formula will match the integrals provided that its order q is high enough. More precisely, since the integrand functions $P_j(\tau)f(\sigma(\tau h))$ have at most degree $\nu s - 1$, all quadrature formulae of order $q \geqslant \nu s$ are capable to solve exactly the integrals and, consequently, they define *equivalent discrete problems* in the form (3.22)–(3.24), which means that the resulting polynomial $u(ch)$ will match $\sigma(ch)$ in (3.10)–(3.12). Thus, for polynomial functions f we get infinitely many equivalent methods.

For the class of HBVM(k, s) methods defined on Gaussian nodes applied to polynomial Hamiltonian systems, we can state the following result that should be read as a completion of Theorem 3.4.

Theorem 3.9. *Let the Hamiltonian function $H(y)$ be a polynomial of degree ν. Then all HBVM(k, s) methods are equivalent for $k \geqslant \nu s/2$.*

As a further example, it is possible to show that the use of a Lobatto quadrature formula (of order $2s$) based at the $k = s + 1$ abscissae $0 = \tau_0 < \tau_1 < \cdots < \tau_s = 1$ yields the Lobatto IIIA collocation method of order $2s$.[9] Therefore, if f is linear the quadrature is exact, and we deduce that all Lobatto IIIA methods and Gauss-Legendre methods sharing the same order are equivalent when applied to linear problems. Consequently, they provide the very same numerical solution.[10]

---

[9]See Corollary 1 in [48].
[10]Up to round-off errors.

3.3.3   *A-stability*

We now show that HBVM(k, s) methods share another relevant property with Gauss-Legendre collocation formulae, namely the *perfect A-stability*.[11]
Given the celebrated *Dahlquist's test equation*,

$$\dot{y} = \lambda y, \qquad y(0) = y_0 \neq 0, \qquad \mathrm{Re}(\lambda) \leqslant 0, \tag{3.40}$$

one has that

$$\lim_{t \to +\infty} y(t) = 0 \Leftrightarrow \mathrm{Re}(\lambda) < 0 \qquad \text{and} \qquad |y(t)| = |y(0)|, \ \forall t \geqslant 0 \Leftrightarrow \mathrm{Re}(\lambda) = 0.$$

Correspondingly, we introduce the following definition regarding the numerical method.

Definition 3.2. *Given a one-step method* $y_1 = \Phi_h(y_0)$ *applied to (3.40) with stepsize* h, *we say that the method is* perfectly *A-stable provided that, for all* $h > 0$ *and* $y_0 \neq 0$:

$$\lim_{n \to \infty} y_n = 0 \Leftrightarrow \mathrm{Re}(\lambda) < 0 \quad \text{and} \quad |\Phi_h(y_0)| = |y_0| \Leftrightarrow \mathrm{Re}(\lambda) = 0.$$

By virtue of the equivalence property discussed in Section 3.3.2, we see that any HBVM(k, s) with $k \geqslant s$ is equivalent to the underlying Gauss-Legendre method of order $2s$ if applied to linear autonomous problems such as (3.40). *A*-stability is then automatically inherited from the Gauss method as well as the stability rational function, that is the (s, s)-Padé approximation to $e^{h\lambda}$. Nevertheless, for completeness we also report a different demonstration of *A*-stability which makes use of the line integral tool: of course, this proof also provides an alternative pathway to show the *A*-stability of the Gauss method.

Theorem 3.10. *For all* $k \geqslant s$, *HBVM(k, s) methods are perfectly A-stable.*

Proof. By setting

$$\lambda = \alpha + \mathrm{i}\beta, \qquad y = x_1 + \mathrm{i}x_2, \qquad x = \begin{pmatrix} x_1 \\ x_2 \end{pmatrix}, \qquad A = \begin{pmatrix} \alpha & -\beta \\ \beta & \alpha \end{pmatrix},$$

with $\alpha, \beta, x_1, x_2 \in \mathbb{R}$, problem (3.40), over the interval $[0, h]$ reads:

$$\dot{x} = Ax \equiv A\nabla V(x), \qquad x(0) = \left(\ \mathrm{Re}(y_0), \ \mathrm{Im}(y_0) \ \right)^\top \neq \mathbf{0}^\top, \tag{3.41}$$

where $V(x) := \frac{1}{2}x^\top x$ is a Lyapunov function for (3.41). Application of the method (3.22)–(3.25) then gives:

$$\dot{u}(ch) = \sum_{j=0}^{s-1} P_j(c)\hat{\gamma}_j, \qquad c \in [0, 1]. \tag{3.42}$$

---

[11]This notion is also referred to as *precise A-stability*.

Consequently, by observing that

$$\hat{\gamma}_j = A \sum_{i=1}^{k} b_i P_j(c_i) \nabla V(u(c_i h)) = A \int_0^1 P_j(c) \nabla V(u(ch)) \mathrm{d}c,$$

since $u \in \Pi_s$ and $P_j \in \Pi_{s-1}$, one obtains:

$$
\begin{aligned}
\Delta V(u(0)) &= V(u(h)) - V(u(0)) = \int_0^h \nabla V(u(t))^\top \dot{u}(t) \mathrm{d}t \\
&= h \int_0^1 \nabla V(u(\tau h))^\top \dot{u}(\tau h) \mathrm{d}\tau = h \int_0^1 \nabla V(u(\tau h))^\top \sum_{j=0}^{s-1} P_j(\tau) \hat{\gamma}_j \mathrm{d}\tau \\
&= h \sum_{j=0}^{s-1} \left[\int_0^1 P_j(\tau) \nabla V(u(\tau h)) \mathrm{d}\tau \right]^\top A \left[\int_0^1 P_j(c) \nabla V(u(ch)) \mathrm{d}c \right] \\
&= \alpha h \sum_{j=0}^{s-1} \left\| \int_0^1 P_j(c) \nabla V(u(ch)) \mathrm{d}c \right\|^2 =: \alpha h \Psi^2.
\end{aligned}
$$

Perfect A-stability then follows, after observing that $\Psi^2 > 0$. In fact, let us assume for a moment $\Psi^2 = 0$. Then, on the one hand we would get $\hat{\gamma}_j = 0$ for $j = 0, \ldots, s-1$ which implies $\dot{u}(ch) \equiv 0$ and, on the other hand, considering that $\nabla V(u) = u$, the s-degree polynomial $u(ch)$ would be orthogonal to all $P_j(c)$, $j = 0, \ldots, s-1$, so $u(ch) = \rho P_s(c)$ for some $\rho \neq 0$. The two conclusions are evidently in contradiction. □

3.3.4 Symmetry

We recall that a generic one-step integrator $y_1 = \Phi_h(y_0)$ is *symmetric* or *time-reversible* if

$$y_1 = \Phi_h(y_0) \quad \Longleftrightarrow \quad y_0 = \Phi_{-h}(y_1), \tag{3.43}$$

or, equivalently,

$$\Phi_h \circ \Phi_{-h} = \text{identity}.$$

In other words, a symmetric method is capable to reproduce the very same numerical solution but in reversed order if applied to the differential equation coupled with the last computed approximation and with the time reversed (hence, with stepsize $-h$).

Symmetry has been recognized as an important ingredient for the good long-time behavior of a numerical method applied to several kind of conservative problems including Hamiltonian problems (see the bibliographical notes).

Basing on (3.43), to check whether a given one-step method is symmetric, one has to first define the reverse map Φ_{-h} which is obtained by reversing the time and replacing the initial state vector y_0 with the final state vector y_1.

Working on a generic time interval $[t_0, t_1]$ of length $h = t_1 - t_0$, this means that

$$
\begin{aligned}
t_0 + ch &\longrightarrow & t_1 - ch, & \quad c \in [0, 1], \\
y_0 &\longrightarrow & y_1,
\end{aligned}
\tag{3.44}
$$

where $y_1 = \Phi_h(y_0)$ is the numerical solution computed by the HBVM(k, s) that we now consistently rewrite on the time interval $[t_0, t_1]$:

$$
y_1 := u(t_0 + h) = y_0 + h \sum_{i=1}^{k} b_i f(u(t_0 + c_i h)) \equiv y_0 + h \hat{\gamma}_0,
\tag{3.45}
$$

with

$$
u(t_0 + ch) = y_0 + h \sum_{j=0}^{s-1} \int_0^c P_j(x) \mathrm{d}x \, \hat{\gamma}_j, \quad c \in [0, 1],
\tag{3.46}
$$

$$
\hat{\gamma}_j = \sum_{i=1}^{k} b_i P_j(c_i) f(u(t_0 + c_i h)), \quad j = 0, \ldots, s - 1.
\tag{3.47}
$$

Theorem 3.11. *For all $k \geqslant s$, the HBVM(k, s) method (3.45), based at the Gauss-Legendre nodes c_i defined at (3.25) is symmetric. Moreover, assuming to apply method (3.45) with a different abscissae distribution, the resulting method remains symmetric under one of the following assumptions:*

(a) the abscissae are symmetrically distributed on $[0, 1]$:

$$
c_i = 1 - c_{k-i+1}, \quad i = 1, \ldots, k,
\tag{3.48}
$$

(b) the function f is a polynomial of degree say μ and the quadrature rule (c_i, b_i) has order at least $q \geqslant (\mu + 1)s$.

Proof. The method Φ_{-h} is defined by introducing the transformations (3.44) in formulae (3.46)-(3.47) yielding the polynomial $\bar{u}$ defined on $[t_0, t_1]$ as

$$
\bar{u}(t_1 - ch) = y_1 - h \sum_{j=0}^{s-1} \int_0^c P_j(x) \mathrm{d}x \, \bar{\gamma}_j, \quad c \in [0, 1],
$$

with the vectors $\bar{\gamma}_j$ satisfying the (nonlinear) system

$$
\begin{aligned}
\bar{\gamma}_j &= \sum_{i=1}^{k} b_i P_j(c_i) f(\bar{u}(t_1 - c_i h)) \\
&\equiv \sum_{i=1}^{k} b_i P_j(c_i) f\left(y_1 - h \sum_{j=0}^{s-1} \int_0^{c_i} P_j(x) \mathrm{d}x \, \bar{\gamma}_j \right), \quad j = 0, \ldots, s - 1.
\end{aligned}
\tag{3.49}
$$

Thus, starting at y_1, the reversed method Φ_{-h} computes the approximation $\bar{y}_0 := \Phi_{-h}(y_1)$ given as

$$
\bar{y}_0 := \bar{u}(t_0) = \bar{u}(t_1 - h) = y_1 - h \bar{\gamma}_0.
\tag{3.50}
$$

We start by proving the symmetry of the methods under assumption *(a)* which also covers the case of a Gauss-Legendre nodes distribution, the default choice in a HBVM(k, s) formula. To this end, it suffices to check that $\bar{y}_0$, defined at (3.50), matches y_0. First we notice that, for interpolation quadrature formulae, (3.48) also implies the symmetry of the weights:

$$b_i = b_{k-i+1}, \qquad i = 1, \ldots, k. \tag{3.51}$$

We then show that the vectors $\bar{\gamma}_j$ and $\hat{\gamma}_j$ defined respectively at (3.49) and (3.47) are related as

$$\bar{\gamma}_j = (-1)^j \hat{\gamma}_j, \qquad j = 0, \ldots, s - 1,$$

which, because of (3.50), evidently implies $\bar{y}_0 = y_0$. Setting $\gamma_j^* := (-1)^j \hat{\gamma}_j$ and exploiting the symmetry property

$$(-1)^j P_j(c) = P_j(1 - c)$$

of Legendre polynomials,[12] from (3.47) and (3.46) we get

$$
\begin{aligned}
\gamma_j^* &= \sum_{i=1}^k b_i (-1)^j P_j(c_i) f\left(y_0 + h \sum_{\ell=0}^{s-1} \int_0^{c_i} P_\ell(x) \mathrm{d}x\, \hat{\gamma}_\ell \right) \\
&= \sum_{i=1}^k b_i P_j(1 - c_i) f\left(y_0 + h \sum_{\ell=0}^{s-1} \left(\int_0^1 P_\ell(x)\mathrm{d}x - \int_{c_i}^1 P_\ell(x)\mathrm{d}x \right) \hat{\gamma}_\ell \right) \\
&= \sum_{i=1}^k b_i P_j(1 - c_i) f\left(y_1 - h \sum_{\ell=0}^{s-1} \int_{c_i}^1 P_\ell(x)\mathrm{d}x\, \hat{\gamma}_\ell \right).
\end{aligned}
$$

Introducing the change of variables $\tau = 1 - x$, the latter integrals become

$$\int_{c_i}^1 P_\ell(x)\mathrm{d}x = -\int_{1-c_i}^0 P_\ell(1 - \tau)\mathrm{d}\tau = (-1)^\ell \int_0^{1-c_i} P_\ell(\tau)\mathrm{d}\tau,$$

and exploiting the symmetry assumptions (3.48) and (3.51) we finally obtain

$$
\begin{aligned}
\gamma_j^* &= \sum_{i=1}^k b_{k+1-i} P_j(c_{k+1-i}) f\left(y_1 - h \sum_{\ell=0}^{s-1} \int_0^{c_{k+1-i}} P_\ell(\tau)\mathrm{d}\tau\, (-1)^\ell \hat{\gamma}_\ell \right) \\
&= \sum_{i=1}^k b_i P_j(c_i) f\left(y_1 - h \sum_{\ell=0}^{s-1} \int_0^{c_i} P_\ell(\tau)\mathrm{d}\tau\, \gamma_\ell^* \right).
\end{aligned}
$$

A comparison with (3.49) shows that the vectors γ_j^* and $\bar{\gamma}_j$ actually satisfy the very same nonlinear system and, therefore, they must coincide.

To deduce the symmetry property under the assumptions listed in item *(b)*, it is sufficient to notice that, in such a case, all the integrals involved in the method are exact. As a consequence of the equivalence property discussed in Section 3.3.2 we conclude that the resulting method is equivalent to a standard HBVM(k', s) method based at Gauss-Legendre nodes, with k' large enough. □

---

[12]See property **P6** in Section A.1.

3.3.5   Link with collocation methods

In this section we wish to better explore the existing connections between HBVMs and Runge-Kutta collocation methods. For this purpose, we now assume that the abscissae distribution of the nodes $c_1 < c_2 < \cdots < c_k \in [0,1]$ be generic. The corresponding collocation method is then defined by the tableau

$$
\begin{array}{c|c}
\begin{array}{c} c_1 \\ \vdots \\ c_k \end{array} & \mathcal{A} \equiv (\alpha_{ij}) \\
\hline
& b_1 \ \ldots \ b_k
\end{array}
\tag{3.52}
$$

where, for $i, j = 1, \ldots, k$:

$$
\alpha_{ij} := \left(\int_0^{c_i} \ell_j(x)\mathrm{d}x \right), \qquad b_j := \int_0^1 \ell_j(x)\mathrm{d}x,
\tag{3.53}
$$

$\ell_j(c)$ being the j-th Lagrange polynomial of degree $k - 1$ defined on the set of abscissae $\{c_i\}$. Given a positive integer $s \leq k$, we now consider the matrix (see (3.34) and (3.35))

$$
\mathcal{P}_s \mathcal{P}_s^\top \Omega \in \mathbb{R}^{k \times k}.
$$

The class of Runge-Kutta methods we are interested in, is then defined by the tableau

$$
\begin{array}{c|c}
\begin{array}{c} c_1 \\ \vdots \\ c_k \end{array} & A := \mathcal{A}(\mathcal{P}_s \mathcal{P}_s^\top \Omega) \\
\hline
& b_1 \ \ldots\ldots \ b_k
\end{array}
\tag{3.54}
$$

We note that the Butcher array A has rank which cannot exceed s, because it is defined by *filtering* $\mathcal{A}$ by the rank s matrix $\mathcal{P}_s \mathcal{P}_s^\top \Omega$. The following result then holds true.

Theorem 3.12. *The tableau (3.54) defines a HBVM(k,s) method based at the abscissae $\{c_i\}$.*

Proof. Let us expand the basis of Legendre polynomials $\{P_0(c), \ldots, P_{s-1}(c)\}$ along the Lagrange basis $\{\ell_j(c)\}$, $j = 1, \ldots, k$, defined over the nodes c_i, $i = 1, \ldots, k$:

$$
P_j(c) = \sum_{r=1}^k P_j(c_r) \ell_r(c), \qquad j = 0, \ldots, s - 1.
$$

It follows that, for $i = 1, \ldots, k$ and $j = 0, \ldots, s - 1$:

$$
\int_0^{c_i} P_j(x)\mathrm{d}x = \sum_{r=1}^k P_j(c_r) \int_0^{c_i} \ell_r(x)\mathrm{d}x = \sum_{r=1}^k P_j(c_r)\alpha_{ir},
$$

that is (see (3.33)–(3.35) and (3.52)-(3.53)), $\mathcal{I}_s = \mathcal{A}\mathcal{P}_s$. Consequently,

$$\mathcal{A}\mathcal{P}_s P_s^\top \Omega = \mathcal{I}_s P_s^\top \Omega, \tag{3.55}$$

so that, in virtue of Theorem 3.6, one retrieves the tableau (3.32) which defines the method HBVM(k, s) defined on the abscissae $\{c_i\}$.   □

From (3.55), considering that $\mathcal{I}_s = \mathcal{P}_{s+1}\hat{X}_s$ (see Lemma 3.5), we see that (3.54) may be cast as (3.39) with the generalized W-transformation $\mathcal{P}_s$ now defined on the abscissae $\{c_i\}$ at hand.

By virtue of the equivalence property stated in Section 3.3.2, if the distribution of the abscissae c_i is generic but f is a polynomial of degree $\nu - 1$, for some $\nu \geqslant 2$, we can still compare the HBVM method (3.54) with a HBVM based at Gaussian nodes. In fact these two methods become equivalent as soon as the order of their underlying quadrature formule is at least νs.

Remark

It would seem that the price paid to achieve the energy-conservation property consists of a reduction of the order of the new method (3.54) with respect to the original one (3.52). For example, if the $\{c_i\}$ are the Gaussian nodes on $[0, 1]$, it is clear that (3.52) has order $2k$ while (3.54) has order $2s$.

Actually, a fair comparison would be to relate method (3.54) to a collocation method constructed on s rather than on k stages. In fact, the actual nonlinear system arising from the implementation of a HBVM(k, s) method turns out to have block dimension s, as a consequence of the fact that rank$(A) = s$. The implementation issues will be thoroughly discussed in the next chapter.

Let us write down the nonlinear system defining the vector $Y := (y_1, \ldots, y_k)^\top$ of the stages of method (3.54) (to simplify the notation, we assume $f(y)$ to be a scalar function):

$$Y = y_0 \mathbf{1} + h\mathcal{A}\mathcal{P}_s P_s^\top \Omega F(Y), \tag{3.56}$$

where $F(Y) := (f(y_1), \ldots, f(y_k))^\top$. Let the order of the quadrature (c_i, b_i) be at least $2s - 1$. This assumption assures that the first two properties listed in Lemma 3.6 remain valid for the generic abscissae distribution we are now considering. In particular we see that, in such a case, matrix $P_s P_s^\top \Omega$ is idempotent and acts as a projector onto the s-dimensional subspace spanned by the columns of $\mathcal{P}_s$. We discuss the two cases:

$s = k$. From the second property listed in Lemma 3.6 we get $P_s^\top \Omega = P_s^{-1}$ so $A \equiv \mathcal{A}$ and method (3.54) becomes the underlying classical collocation method (3.52).

$s < k$. Formula (3.56) may be interpreted as a generalization of the definition of a standard collocation method in the following sense: matrix $\mathcal{A}$ is applied to the vector $G(Y) := \mathcal{P}_s\mathcal{P}_s^\top\Omega F(Y)$ instead of $F(Y)$. What does this mean? The vector $F(Y)$ is projected onto the subspace spanned by the columns of $\mathcal{P}_s$, which consists of vectors containing the evaluations of the Legendre polynomials over the mesh $\{c_i\}$. This process is precisely the discrete counterpart of the procedure discussed at the beginning of the chapter to derive the methods: projection of the function $f(y(ch))$ onto the subspace Π_{s-1} and replacement of the original problem $\dot{y}(ch) = f(y(ch))$ with the projected one $\dot{\sigma}(ch) = \sum_{i=0}^{s-1} P_i(c)\int_0^1 P_i(\tau)f(\sigma(\tau h))\mathrm{d}\tau$. Summarizing, the case $k > s$ may be interpreted as applying the standard collocation method (3.52) to a projected vector function.

3.4   Least square approximation and Fourier expansion

Our starting point here is the following remark concerning the local truncated problem (3.11), namely $\dot{\sigma}(ch) = \sum_{j=0}^{s-1} P_j(c)\gamma_j$.[13] A comparison with (3.9) shows that the residual $\dot{\sigma}(ch) - f(\sigma(ch))$ satisfies [14]

$$\dot{\sigma}(ch) - f(\sigma(ch)) = \sum_{j \geqslant s} P_j(c)\gamma_j.$$

Since $\gamma_j = \int_0^1 P_j(\tau)f(\sigma(\tau h))\mathrm{d}\tau$ is the projection of $f(\sigma(ch))$ along the j-th polynomial of the (complete orthogonal) Legendre basis, $\dot{\sigma}(ch)$ turns out to be the projection of the function $f(\sigma(\tau h))$ onto the s-dimensional subspace Π_{s-1} spanned by the polynomials $P_j(ch)$, $j = 0, \ldots, s-1$. As a consequence, the residual $\dot{\sigma}(ch) - f(\sigma(ch))$ is orthogonal to Π_{s-1}. This is reminiscent of the classical approach to define the best approximant to a given element belonging to an infinite-dimensional Hilbert space from a given subspace.

However, in the present context, one should be warned that the projection is implicit in nature, since the coefficients γ_j also appear in the object we wish to approximate, namely $f(\sigma(ch)) = f(y_0 + h\sum_{i=0}^{s-1}\int_0^c P_i(x)\mathrm{d}x\,\gamma_j)$. As a result, we cannot conclude that $\sigma(ch)$ actually minimizes the L_2 norm of the residual among all polynomials $q(ch) \in \Pi_{s-1}$ satisfying $q(0) = y_0$. In principle, a slight perturbation introduced in $\dot{\sigma}(ch)$ could cause $f(\sigma(ch))$ to get close enough to Π_{s-1} that the resulting residual improves on the one corresponding to σ. Consequently, in this section we want to explore the extent to which the polynomial σ is related to the least-square fitting polynomial. For this

---

[13] For sake of brevity, we shall use the notation γ_j in place of $\gamma_j(\sigma)$.

[14] Such a property has already been used in the proof of Theorem 3.3.

purpose, we consider a general non-autonomous IVP:

$$\begin{cases} \dot{y}(ch) = f(ch, y(ch)), & c \in [0, 1], \\ y(0) = y_0. \end{cases} \tag{3.57}$$

To simplify the notation, but without loss of generality, we assume f to be a scalar function.

As was clear from the arguments exposed in Section 3.1, confining the study on the time window $[0, h]$ has the following sense: we first approximate (3.57) by a new IVP, say $\dot{\sigma}(ch) = g(ch, \sigma(ch))$ with $\sigma(0) = y_0$ and $g \approx f$, whose solution $\sigma(ch)$ may be analytically determined; then we use $\sigma(t)$ as an approximation to $y(t)$ for $t \in [0, h]$. Thus, the simplified IVP can be thought of as a numerical integrator for which the parameter $h > 0$ stands for the stepsize of integration. This approach is typical for deriving standard collocation methods and has been used throughout this chapter to devise line integral methods.

Here, in the same spirit as for collocation and line integral methods, we wish to approximate the solution $y(t)$, for $t = ch \in [0, h]$, of problem (3.57), by means of a polynomial of degree s, say $q(t)$, such that $\dot{q}(t)$ is "close" to $f(t, q(t))$ on $[0, h]$. More precisely, we now try to minimize the residual $\dot{q}(t) - f(t, q(t))$ in a least square sense. Consequently, we search for the polynomial $\bar{\sigma} \in \Pi_s$, with $\bar{\sigma}(0) = y_0$, such that

$$\int_0^1 \left(\dot{\bar{\sigma}}(ch) - f(ch, \bar{\sigma}(ch)) \right)^2 dc = \min_{q \in \Pi_s, \ q(0) = y_0} \int_0^1 \left(\dot{q}(ch) - f(ch, q(ch)) \right)^2 dc. \tag{3.58}$$

Remark

We observe that, by approximating the integrals in (3.58) by a polynomial interpolatory formula, based at s abscissae $c_1, \ldots, c_s$ and with *positive* weights $b_1, \ldots, b_s$, provides us with the *collocation polynomial* at such abscissae:

$$\dot{\bar{\sigma}}(c_i h) = f(c_i h, \bar{\sigma}(c_i h)), \quad i = 1, \ldots, s, \qquad \bar{\sigma}(0) = y_0.$$

In fact, in such a case, one has:

$$\sum_{i=1}^{s} b_i \left(\dot{\bar{\sigma}}(c_i h) - f(c_i h, \bar{\sigma}(c_i h)) \right)^2 = 0.$$

The reason why, in the present context, we are interested in problem (3.58) is that the least square fitting polynomial $\bar{\sigma}$ is intimately related to the polynomial σ defining the line integral method and our purpose here is to inspect this relationship. For convenience, we write again the IVP defining the polynomial σ on the interval $[0, h]$:

$$\begin{cases} \dot{\sigma}(ch) = \displaystyle\sum_{j=0}^{s-1} P_j(c)\gamma_j, \quad \text{with } \gamma_j = \int_0^1 P_j(\tau)f(\tau h, \sigma(\tau h))\mathrm{d}\tau, \\ \sigma(0) = y_0. \end{cases} \tag{3.59}$$

We now specify an arbitrary polynomial $q(ch) \in \Pi_s$ by expanding its derivative along the Legendre basis, namely

$$\dot{q}(ch) = \sum_{j=0}^{s-1} P_j(c)\lambda_j, \tag{3.60}$$

where λ_j are arbitrary real coefficients. Consequently, imposing the condition $q(0) = y_0$ yields

$$q(ch) = y_0 + h \sum_{j=0}^{s-1} \int_0^c P_j(x)\mathrm{d}x\lambda_j. \tag{3.61}$$

For later use, we also consider the expansion of $f(ch, q(ch))$

$$f(ch, q(ch)) = \sum_{j\geq 0} P_j(c)\eta_j, \quad \text{with } \eta_j = \int_0^1 P_j(\tau)f(\tau h, q(\tau h))\mathrm{d}\tau, \tag{3.62}$$

and notice that choosing $\lambda_j = \eta_j$, $j = 0, \dots, s-1$, leads us back to system (3.14): thus, this choice is tantamount to set $\lambda_j = \gamma_j$ (compare (3.61)-(3.62) with (3.59)) or, in other words, $q(ch) = \sigma(ch)$.

In order to derive the least square polynomial approximation $\bar{\sigma}$, we determine the critical points of the function

$$F(\lambda_0, \dots, \lambda_{s-1}) = \int_0^1 (\dot{q}(ch) - f(ch, q(ch)))^2\,\mathrm{d}c$$

$$= \int_0^1 \left(\sum_{j=0}^{s-1} P_j(c)\lambda_j - f\left(ch, y_0 + h\sum_{j=0}^{s-1}\int_0^c P_j(x)\mathrm{d}x\lambda_j\right)\right)^2\,\mathrm{d}c.$$

It turns out that

$$\frac{\partial F}{\partial \lambda_k} = 2\int_0^1 (\dot{q}(ch) - f(ch, q(ch)))\left(P_k(c) - h\frac{\partial f}{\partial y}(ch, q(ch))\int_0^c P_k(x)\mathrm{d}x\right)\mathrm{d}c.$$

Thus, upon observing that, by virtue of (3.60),

$$\int_0^1 P_k(c)\dot{q}(ch)\mathrm{d}c = \lambda_k,$$

the stationary points of F are the solutions of the nonlinear system

$$\lambda_k - \int_0^1 P_k(c)f(ch, q(ch))\mathrm{d}c = \int_0^1 (\dot{q}(ch) - f(ch, q(ch)))\, g_k(ch)\mathrm{d}c,$$

$$k = 0, \dots, s-1, \tag{3.63}$$

with

$$g_k(ch) = h\frac{\partial f}{\partial y}(ch, q(ch)) \int_0^c P_k(x)\mathrm{d}x = \frac{\partial f}{\partial y}(ch, q(ch)) \int_0^{ch} P_k\left(\frac{x}{h}\right) \mathrm{d}x. \quad (3.64)$$

Theorem 3.13. *Under regularity assumptions on the function f, system (3.63) admits a unique solution $(\bar{\lambda}_0, \ldots, \bar{\lambda}_{s-1})^\top$ for all sufficiently small h. Furthermore, the least square fitting polynomial*

$$\bar{\sigma}(ch) = y_0 + h\sum_{j=0}^{s-1} \int_0^c P_j(x)\mathrm{d}x\,\bar{\lambda}_j \qquad (3.65)$$

and the polynomial $\sigma(ch)$ defining a line integral method (see (3.59)) are related as follows:

(a) $\bar{\sigma}(ch) = \sigma(ch) + O(h^{2s+1})$;

(b) the least square errors

$$\int_0^1 \left(\dot{\bar{\sigma}}(ch) - f(ch, \bar{\sigma}(ch))\right)^2 \mathrm{d}c \quad and \quad \int_0^1 \left(\dot{\sigma}(ch) - f(ch, \sigma(ch))\right)^2 \mathrm{d}c$$

are both $O(h^{2s})$ as $h \to 0$ and their expansion in power of the stepsize h coincide up to the order $4s - 1$, namely:

$$\int_0^1 \left(\dot{\bar{\sigma}}(ch) - f(ch, \bar{\sigma}(ch))\right)^2 \mathrm{d}c - \int_0^1 \left(\dot{\sigma}(ch) - f(ch, \sigma(ch))\right)^2 \mathrm{d}c = O(h^{4s}).$$

Proof. In view of (3.59), substituting γ_k in place of λ_k in (3.63), and hence $\sigma(ch)$ in place of $q(ch)$, makes the left-hand side vanish. Concerning the right-hand side, exploiting Lemmas 3.1 and 3.2, we have

$$\int_0^1 \left(\dot{\sigma}(ch) - f(ch, \sigma(ch))\right) g_k(ch)\mathrm{d}c = -\int_0^1 \sum_{j\geqslant s} P_j(c)\gamma_j\, g_k(ch)\mathrm{d}c$$

$$= -\sum_{j\geqslant s} \gamma_j \int_0^1 P_j(c) g_k(ch)\mathrm{d}c = O(h^{2s}). \qquad (3.66)$$

Thus, in a neighborhood of $\gamma = \left(\begin{array}{ccc}\gamma_0, & \ldots, & \gamma_{s-1}\end{array}\right)^\top$ (see (3.16)), system (3.63) may be thought of as a perturbation of system (3.14), the perturbing term being the right-hand side. This suggests that (3.63) may possess a solution $\bar{\lambda} = \left(\begin{array}{ccc}\bar{\lambda}_0, & \ldots, & \bar{\lambda}_{s-1}\end{array}\right)^\top$ very close to γ. To see that this is indeed the case, and to also derive the statement *(a)*, we recast system (3.63) in vector form as

$$\lambda = \Psi(\lambda) + R(\lambda),$$

where

$$\boldsymbol{\lambda} := \begin{pmatrix} \lambda_0 \\ \vdots \\ \lambda_{s-1} \end{pmatrix}, \qquad \Psi(\boldsymbol{\lambda}) := \begin{pmatrix} \int_0^1 P_0(c) f(ch, q(ch)) \mathrm{d}c \\ \vdots \\ \int_0^1 P_{s-1}(c) f(ch, q(ch)) \mathrm{d}c \end{pmatrix},$$

and

$$R(\boldsymbol{\lambda}) := \begin{pmatrix} \int_0^1 \left(\dot{q}(ch) - f(ch, q(ch)) \right) g_0(ch) \mathrm{d}c \\ \vdots \\ \int_0^1 \left(\dot{q}(ch) - f(ch, q(ch)) \right) g_{s-1}(ch) \mathrm{d}c \end{pmatrix}.$$

We notice that $\Psi(\boldsymbol{\lambda})$ is the same function we introduced in the proof of Theorem 3.1 at (3.16). Thus, it is Lipschitz continuous with constant $K = O(h)$, according to (3.17), and its fixed point is $\boldsymbol{\gamma}$. A direct computation shows that $\partial R(\boldsymbol{\lambda})_k / \partial \lambda_i = O(h)$, which implies that $R(\boldsymbol{\lambda})$ is Lipschitz continuous as well, with constant $L = O(h)$. In fact, from (3.64) we see that $g_k(ch) = O(h)$ and

$$\frac{\partial g_k}{\partial \lambda_i}(ch) = h^2 \frac{\partial^2 f}{\partial y^2}(ch, q(ch)) \int_0^c P_i(x) \mathrm{d}x \int_0^c P_k(x) \mathrm{d}x = O(h^2).$$

Thus the function $\Psi(\boldsymbol{\lambda}) + R(\boldsymbol{\lambda})$ is Lipschitz with constant $\nu \leqslant K + L = O(h)$ and is, therefore, a contraction for h small enough: consequently, according to the contraction mapping theorem, it admits a unique fixed point $\bar{\boldsymbol{\lambda}}$. In order to locate it, we consider the closed ball $B(\boldsymbol{\gamma}, \rho) \subset \mathbb{R}^s$ with center $\boldsymbol{\gamma}$ and radius $\rho = \|R(\boldsymbol{\gamma})\|/(1 - \nu)$ and notice that (3.66) yields $\rho = O(h^{2s})$. To deduce that the fixed point $\bar{\boldsymbol{\lambda}} \in B(\boldsymbol{\gamma}, \rho)$, it suffices to show that the sequence $\{\boldsymbol{\lambda}_k\}$, defined as

$$\begin{cases} \boldsymbol{\lambda}_0 = \boldsymbol{\gamma}, \\ \boldsymbol{\lambda}_{k+1} = \Psi(\boldsymbol{\lambda}_k) + R(\boldsymbol{\lambda}_k), \qquad k = 0, 1, \ldots, \end{cases} \tag{3.67}$$

is entirely contained in the closed ball $B(\boldsymbol{\gamma}, \rho)$. We proceed by induction. Starting with $k = 1$ we get

$$\|\boldsymbol{\lambda}_1 - \boldsymbol{\gamma}\| = \|\boldsymbol{\lambda}_1 - \Psi(\boldsymbol{\gamma})\| = \|R(\boldsymbol{\gamma})\| \equiv (1 - \nu)\rho.$$

Assuming now that $\boldsymbol{\lambda}_k \in B(\boldsymbol{\gamma}, \rho)$, we obtain

$$\begin{aligned} \|\boldsymbol{\lambda}_{k+1} - \boldsymbol{\gamma}\| &\leqslant \|\boldsymbol{\lambda}_{k+1} - \boldsymbol{\lambda}_1\| + \|\boldsymbol{\lambda}_1 - \boldsymbol{\gamma}\| \\ &= \|(\Psi + R)(\boldsymbol{\lambda}_k) - (\Psi + R)(\boldsymbol{\gamma})\| + (1 - \nu)\rho \\ &\leqslant \nu \|\boldsymbol{\lambda}_k - \boldsymbol{\gamma}\| + (1 - \nu)\rho \\ &\leqslant \nu\rho + (1 - \nu)\rho = \rho. \end{aligned}$$

In conclusion, for h sufficiently small, the iteration (3.67) produces a convergent sequence $\{\boldsymbol{\lambda}_k\}$, whose limit $\bar{\boldsymbol{\lambda}}$ is the unique solution of system (3.63) and satisfies

$$\|\bar{\boldsymbol{\lambda}} - \boldsymbol{\gamma}\| = O(h^{2s}). \tag{3.68}$$

Consequently, statement *(a)* follows, since:

$$r(ch) := \bar{\sigma}(ch) - \sigma(ch) = h \sum_{j=0}^{s-1} \int_0^c P_j(x)\mathrm{d}x \, (\bar{\lambda}_j - \gamma_j) = O(h^{2s+1}). \quad (3.69)$$

To prove *(b)* we first recall that

$$\dot{\sigma}(ch) - f(ch, \sigma(ch)) = \sum_{j=0}^{s-1} P_j(c)\gamma_j - \sum_{j=0}^{\infty} P_j(c)\gamma_j = -\sum_{j=s}^{\infty} P_j(c)\gamma_j = O(h^s),$$

and, consequently,

$$\int_0^1 (\dot{\sigma}(ch) - f(ch, \sigma(ch)))^2 \, \mathrm{d}c = O(h^{2s}).$$

Moreover, from (3.69) one obtains:

$$\dot{r}(ch) = \dot{\bar{\sigma}}(ch) - \dot{\sigma}(ch) = \sum_{j=0}^{s-1} P_j(c) \, (\bar{\lambda}_j - \gamma_j) = O(h^{2s}),$$

and, evidently,

$$\int_0^1 (\dot{\sigma}(ch) - f(ch, \sigma(ch))) \, \dot{r}(ch)\mathrm{d}c$$

$$= -\int_0^1 \left(\sum_{j \geqslant s} P_j(c)\gamma_i \right) \left(\sum_{j=0}^{s-1} P_j(c) \, (\bar{\lambda}_j - \gamma_j) \right) \mathrm{d}c = 0.$$

As a consequence, from the last two equations, one has:

$$\int_0^1 (\dot{\bar{\sigma}}(ch) - f(ch, \bar{\sigma}(ch)))^2 \, \mathrm{d}c$$

$$= \int_0^1 (\dot{\sigma}(ch) + \dot{r}(ch) - f(ch, \sigma(ch) + r(ch)))^2 \, \mathrm{d}c$$

$$= \int_0^1 \Big(\dot{\sigma}(ch) - f(ch, \sigma(ch)) + \dot{r}(ch) \underbrace{- \frac{\partial f}{\partial y}(ch, \sigma(ch))r(ch) + O(h^{4s+2})}_{=: \rho(ch)h^{2s+1}} \Big)^2 \mathrm{d}c$$

$$= \int_0^1 \left(\dot{\sigma}(ch) - f(ch, \sigma(ch)) + \dot{r}(ch) + \rho(ch)h^{2s+1} \right)^2 \, \mathrm{d}c$$

$$= \int_0^1 (\dot{\sigma}(ch) - f(ch, \sigma(ch)))^2 \, \mathrm{d}c + \overbrace{\int_0^1 \dot{r}(ch)^2 \mathrm{d}c}^{=O(h^{4s})}$$

$$+ \, 2h^{2s+1} \int_0^1 (\dot{\sigma}(ch) - f(ch, \sigma(ch))) \, \rho(ch)\mathrm{d}c + O(h^{4s+1})$$

$$= \int_0^1 (\dot\sigma(ch) - f(ch, \sigma(ch)))^2 \, dc \; + \; O(h^{4s}) \; - \; 2h^{2s+1} \underbrace{\sum_{j \geqslant s} \gamma_j \int_0^1 P_j(c)\rho(ch)dc}_{=O(h^{2j})}$$

$$= \int_0^1 (\dot\sigma(ch) - f(ch, \sigma(ch)))^2 \, dc \; + \; O(h^{4s}).$$

This completes the proof. $\qquad\qquad\qquad\qquad\qquad\qquad\qquad\qquad\qquad\qquad\qquad$ $\square$

Remarks

- The proof of Theorem 3.13 suggests a natural way to compute $\bar\lambda$ and, hence, $\bar\sigma$. One starts from the initial guess $\lambda_0 = \gamma$ and then constructs the sequence (3.67). Since $R(\gamma) = O(h^{2s})$, the sequence will converge to $\bar\lambda$ in a few iterates for h small. This algorithm has been used to obtain the results listed in Table 3.1.

- From *(a)* we deduce that the method defined by the least square fitting polynomial, through $y_1 = \bar\sigma(h)$, has still order $2s$. Moreover, its approximation to the true solution inside the interval $[0, h]$ is of the same order as that provided by the polynomial $\sigma(ch)$.

- Of course, the residual $\dot{\bar\sigma}(ch) - f(ch, \bar\sigma(ch))$ is no longer orthogonal to Π_{s-1} unless $\sigma = \bar\sigma$: this is the case, for example, of pure quadrature problems (for which $\partial f/\partial y = 0$).

In order to assess the result of Theorem 3.13, let us consider the following problem:

$$\dot y(ch) = e^{y(ch)} =: f(y(ch)), \qquad c \in [0,1], \qquad y(0) = 1. \qquad (3.70)$$

Let us then compute the polynomials, both of degree $s = 1$, σ_h as in (3.59) and the least square approximation $\bar\sigma_h$ defined at (3.65),[15] and define

$$\phi_h := \|\dot\sigma_h - f(\sigma_h)\|_h^2, \qquad \psi_h := \|\dot{\bar\sigma}_h - f(\bar\sigma_h)\|_h^2, \qquad (3.71)$$

where

$$\|g\|_h^2 = \int_0^1 g(ch)^2 dc. \qquad (3.72)$$

According to the result of Theorem 3.13, we should obtain:

$$\phi_h = O(h^2), \quad \psi_h = O(h^2), \quad \|\sigma_h - \bar\sigma_h\|_h = O(h^3), \quad \phi_h - \psi_h = O(h^4) \geqslant 0. \qquad (3.73)$$

These facts are clearly confirmed by the results listed in Table 3.1.[16]

---

[15] We have used the suffix h, to emphasize the dependence on such a parameter.

[16] For sake of brevity, we have not listed the data corresponding to (3.68), which are also confirmed.

TABLE 3.1: Obtained results for (3.70)-(3.71), by using the norm (3.72), confirming (3.73).

h	ϕ_h	rate	ψ_h	rate	$\|\sigma_h - \bar{\sigma}_h\|_h$	rate	$\phi_h - \psi_h$	rate
10^{-1}	8.770e-02	–	8.626e-02	–	2.611e-03	–	1.445e-03	–
$2^{-1}10^{-1}$	1.529e-02	2.52	1.526e-02	2.50	1.889e-04	3.79	3.673e-05	5.30
$2^{-2}10^{-1}$	3.276e-03	2.22	3.275e-03	2.22	1.869e-05	3.34	1.559e-06	4.56
$2^{-3}10^{-1}$	7.620e-04	2.10	7.619e-04	2.10	2.095e-06	3.16	8.135e-08	4.26
$2^{-4}10^{-1}$	1.839e-04	2.05	1.839e-04	2.05	2.484e-07	3.08	4.658e-09	4.13
$2^{-5}10^{-1}$	4.520e-05	2.02	4.520e-05	2.02	3.026e-08	3.04	2.788e-10	4.06
$2^{-6}10^{-1}$	1.120e-05	2.01	1.120e-05	2.01	3.734e-09	3.02	1.706e-11	4.03
$2^{-7}10^{-1}$	2.789e-06	2.01	2.789e-06	2.01	4.638e-10	3.01	1.055e-12	4.02
$2^{-8}10^{-1}$	6.957e-07	2.00	6.957e-07	2.00	5.778e-11	3.00	6.558e-14	4.01
$2^{-9}10^{-1}$	1.738e-07	2.00	1.738e-07	2.00	7.212e-12	3.00	4.088e-15	4.00

3.5   Related approaches

Hereafter, we briefly discuss some well-known and independent approaches that have led to energy-conserving methods equivalent or intimately related to the class of HBVMs. We consider again the *original problem*

$$\dot{y}(ch) = f(y(ch)), \qquad c \in [0,1], \qquad y(0) = y_0, \tag{3.74}$$

the *projected problem*, that is the approximation to (3.74) into the space Π_{s-1},

$$\dot{\sigma}(ch) = \sum_{j=0}^{s-1} P_j(c)\gamma_j(\sigma), \qquad c \in [0,1], \qquad \sigma(0) = y_0,$$
$$\gamma_j(\sigma) = \eta_j \int_0^1 P_j(\tau)f(\sigma(\tau h))\mathrm{d}\tau, \qquad j = 0, \ldots, s-1, \tag{3.75}$$

where we have used again the parameters η_j introduced in Section 1.6 (see, e.g., (1.77) on page 36), and the final *full discrete problem* (HBVM) (compare with (1.97)-(1.98)),

$$\dot{u}(ch) = \sum_{j=0}^{s-1} P_j(c)\hat{\gamma}_j, \qquad c \in [0,1], \qquad u(0) = y_0,$$
$$\hat{\gamma}_j = \eta_j \sum_{i=1}^{k} b_i P_j(c_i)f(u(c_i h)), \qquad j = 0, \ldots, s-1, \tag{3.76}$$

obtained by approximating the integrals appearing in (3.75) by means of interpolatory quadrature rules. The line integral methods (3.76) have been derived in Section 1.6 (see, e.g., (1.73)–(1.75) on page 35) by exploiting the properties of the discrete line integral, so here $\{P_j(c)\}$ denotes an arbitrary basis of Π_{s-1}

and the constants η_j are determined by imposing the order conditions, as has been shown in Section 1.6.6 on page 44. Consequently, these formulae bring the most general instance of a line integral method of Runge-Kutta type. In the present context, these formulae turn out to be advantageous to better elucidate the relationship between other kind of well-known and independent routes of investigations.

The polynomial approximations $\sigma(ch)$ and $u(ch)$ are obtained by integrating the above equations with respect to the variable c. For example, from (3.75) we get:

$$\sigma(ch) = y_0 + h \sum_{j=0}^{s-1} \eta_j \int_0^c P_j(x)\mathrm{d}x \int_0^1 P_j(\tau)f(\sigma(\tau h))\mathrm{d}\tau, \qquad (3.77)$$

so, the numerical approximation at time $t = h$ is

$$y_1 := \sigma(h) = y_0 + h \sum_{j=0}^{s-1} \eta_j \int_0^1 P_j(x)\mathrm{d}x \int_0^1 P_j(\tau)f(\sigma(\tau h))\mathrm{d}\tau. \qquad (3.78)$$

We notice that (3.77) together with (3.78) may be interpreted as Runge-Kutta method with continuous stage $c \in [0,1]$, which becomes a line integral Runge-Kutta method after introducing an interpolatory quadrature formula to handle the integrals.

As we have often emphasized, the projected and the full discretized problems are the same when f is a polynomial and the order of quadrature is high enough, so $\sigma(ch) \equiv u(ch)$ in this case. The analogy between the two problems may be also extended in the non-polynomial case, taking into account that:

- in finite precision arithmetic, there is no practical difference between solving the integrals exactly or approximating them to within machine precision by means of a quadrature formula of suitable order;

- increasing the order of the quadrature does not affect the dimension, s, of the discrete problem generated by the resulting method and, from a theoretical viewpoint, taking the limit as $k \to \infty$ in (3.76) leads back to problem (3.75).

AVF (Averaged Vector Field) method

With the aim of conserving the energy of a canonical Hamiltonian system, Quispel and McLaren proposed the following formula[17]

$$\frac{y_1 - y_0}{h} = \int_0^1 f((1-\tau)y_0 + \tau y_1))\mathrm{d}\tau, \qquad (3.79)$$

---

[17]As an example of discrete gradient, it appeared in a previous work by Mc Lachlan et al. [132].

that was called *averaged vector field method* since the right-hand side may be interpreted as the integral mean of the vector field $f(y)$ evaluated along the segment joining y_0 to y_1. To prove the energy-conservation property when $f(y) = J\nabla H(y)$, it is sufficient to multiply both sides of the equation by $(\int_0^1 \nabla H((1-\tau)y_0+\tau y_1))d\tau)^\top$. Upon observing that the resulting scalar product at the right-hand side vanishes (since J is skew-symmetric), we get

$$\frac{1}{h}\int_0^1 (y_1 - y_0)^\top \nabla H((1-\tau)y_0 + \tau y_1))d\tau = 0,$$

that is, the line integral of $\nabla H(y)$ along the path $\sigma(\tau h) = (1-\tau)y_0+\tau y_1$, which matches the variation of $H(y)$ at the end points. This was indeed the flow of computations the authors proposed, even though they instead interpreted the latter integral in terms of the fundamental theorem of calculus:

$$\begin{aligned} 0 &= \frac{1}{h}\int_0^1 (y_1 - y_0)^\top \nabla H((1 - \tau)y_0 + \tau y_1))d\tau \\ &= \frac{1}{h}\int_0^1 \frac{d}{d\tau}H((1 - \tau)y_0 + \tau y_1))d\tau \ = \ \frac{1}{h}(H(y_1) - H(y_0)). \end{aligned}$$

Thus formula (3.79) belongs to the class of line integral methods (3.78) by setting $\sigma(\tau h) = (1 - \tau)y_0 + \tau y_1$ (thus $s = 1$) and $\eta_0 = 1.$[18] It was derived in Section 1.5 and its discretization (1.66) has led to the definition of a family of formulae called *s*-stage trapezoidal methods since, when only the two interpolation points $c_1 = 0$ and $c_2 = 1$ are used, we get the standard trapezoidal method. This latter class of Runge-Kutta methods has been also independently derived by Celledoni et al. in a subsequent work, where it was shown that the AVF and the *s*-stage trapezoidal methods are indeed B-series methods of order two [68].

Energy-preserving collocation methods

Hairer introduced these formulae as a generalization to high orders of the AVF method and to compute the limit of HBVM(k, s) methods, as $k \rightarrow \infty$. Given an interpolatory quadrature formula defined on a set of distinct abscissae $c_1, c_2, \ldots, c_s \in [0, 1]$ and with weights $b_i \neq 0$, an approximation to the solution of $\dot{y}(t) = f(y(t))$ on the interval $[0, h]$ is obtained by a polynomial $\sigma(ch)$ defined by means of the following collocation-like conditions:

$$\dot{\sigma}(c_i h) = \frac{1}{b_i}\int_0^1 \ell_i(\tau)f(\sigma(\tau h))d\tau, \quad i = 1, \ldots, s, \quad \sigma(0) = y_0, \qquad (3.80)$$

where $\ell_i(\tau)$ denote the Lagrange polynomials defined on the abscissae c_i. We see that the projected problem (3.75) includes formulae (3.80) by choosing

---

[18]More precisely the AVF method is the HBVM$(\infty, 1)$ method defined at (3.30).

$P_{j-1}(\tau) := \ell_j(\tau)$. In fact, setting $f(y) \equiv 1$, and considering that $\int_0^1 \ell_i(\tau)\mathrm{d}\tau = b_i$, yield $\eta_{i-1} = 1/b_i$.

Denoting by q the order of the quadrature formula (c_i, b_i) and by p the order of formula (3.80), and interpreting the energy-preserving method as a Runge-Kutta method with a continuum of stages $\tau \in [0,1]$ (the analogue of (3.77)), Hairer proved that

$$p = \begin{cases} 2s, & \text{for } q \geqslant 2s - 1, \\ 2q - 2s + 2, & \text{for } q \leqslant 2s - 2. \end{cases}$$

The order is maximized if, for example, the abscissae c_i are located at the s Gaussian nodes on $[0,1]$. In such a case, it is not difficult to verify that the polynomial $\sigma(ch)$ defined by a HBVM(∞, s) (see (3.30) on page 94) matches the polynomial defined in (3.80). In fact, choosing now P_j as the $(j+1)$-st shifted and normalized Legendre polynomial and considering that

$$\frac{1}{b_i}\int_0^1 P_j(x)\ell_i(x)\mathrm{d}x = \frac{1}{b_i}\sum_{r=1}^{s} b_r P_j(c_r)\ell_i(c_r) = P_j(c_i), \qquad j = 0, \ldots, s-1,$$

from (3.10)-(3.11) on page 84, we obtain:

$$\begin{aligned}
\dot{\sigma}(c_i h) &= \sum_{j=0}^{s-1} P_j(c_i) \int_0^1 P_j(\tau) f(\sigma(\tau h))\mathrm{d}\tau \\
&= \sum_{j=0}^{s-1} \left(\frac{1}{b_i}\int_0^1 P_j(x)\ell_i(x)\mathrm{d}x\right) \int_0^1 P_j(\tau) f(\sigma(\tau h))\mathrm{d}\tau \\
&= \frac{1}{b_i}\int_0^1 \underbrace{\left(\sum_{j=0}^{s-1} P_j(\tau) \int_0^1 P_j(x)\ell_i(x)\mathrm{d}x\right)}_{=\ell_i(\tau)} f(\sigma(\tau h))\mathrm{d}\tau \\
&= \frac{1}{b_i}\int_0^1 \ell_i(\tau) f(\sigma(\tau h))\mathrm{d}\tau.
\end{aligned}$$

More in general, whatever the choice of the abscissae c_i, the use of any interpolatory quadrature formula applied to solve the integrals transforms (3.80) into an HBVM method of the form (3.76).

An interesting study concerning the property of conjugate symplecticity has revealed that energy-preserving collocation methods of maximal order $2s$ are conjugate-symplectic up to order $2s + 2$, but not up to a higher order.[19]

Time-finite element methods

A further interesting approach is based on the concept of time-finite element methods, where one finds local Galerkin approximations on each time

---

[19]See Section 1.4, and in particular page 21, for the definition and the meaning of conjugate-symplecticity.

subinterval of a given mesh of size h. To sketch the idea of the time-finite element approach, we focus on the *continuous time-finite element (C-TFE) methods* since they provide the energy-conservation property and turn out to be closely related to the class of HBVMs.[20] With reference to (3.74), the corresponding Galerkin variational problem is formally defined as follows:

Find a polynomial $\sigma(ch) \in \Pi_s$ such that the following condition holds true for arbitrary $v(ch) \in \Pi_{s-1}$:

$$\begin{cases} \int_0^1 v(\tau h)^\top (\dot\sigma(\tau h) - f(\sigma(\tau h)))\mathrm{d}\tau = 0, \\ \sigma(0) = y_0. \end{cases} \tag{3.81}$$

The implementation of TFE methods works similarly to a standard one-step method, in that these methods consist of a sequence of variational problems defined locally on contiguous time elements of length h and which are to be solved sequentially. For example, at the very first step, one sets $y_1 := \sigma(h)$.

Taking $f(y) = J\nabla H(y)$, and choosing $v(ch) = J^\top \dot\sigma(ch)$ yields the energy-conservation property at the end points of the polynomial curve $\sigma(ch)$:

$$\begin{aligned} 0 &= \int_0^1 \left(\overbrace{\dot\sigma(\tau h)^\top J \dot\sigma(\tau h)}^{=0} - \dot\sigma(\tau h)^\top \overbrace{J^2}^{=-I} \nabla H(\sigma(\tau h)) \right) \mathrm{d}\tau \\ &= \int_0^1 \dot\sigma(\tau h)^\top \nabla H(\sigma(\tau h))\mathrm{d}\tau \\ &= H(\sigma(h)) - H(\sigma(0)) = H(y_1) - H(y_0). \end{aligned}$$

Notice that (3.81) means that we are projecting the original problem (3.74) onto Π_{s-1}, the subspace of polynomial of degree at most $s-1$. This is nothing but the result of the truncated Fourier expansion analyzed in the present chapter (compare with the discussion presented at the beginning of Section 3.4). Thus, it is not surprising that the polynomial $\sigma(ch)$ solution of (3.81) is the same as the polynomial associated with the HBVM(∞, s) method of order $2s$. To see this in more detail, we first observe that problem (3.81) is equivalent to

$$\begin{cases} \int_0^1 P_j(\tau)(\dot\sigma(\tau h) - f(\sigma(\tau h)))\mathrm{d}\tau = 0, \qquad j = 0, \ldots, s-1, \\ \sigma(0) = y_0, \end{cases} \tag{3.82}$$

where P_j denotes the j-th shifted and normalized Legendre polynomial on $[0,1]$. Expanding $\dot\sigma(ch)$ along this basis

$$\dot\sigma(\tau h) = \sum_{i=0}^{s-1} P_i(\tau)\gamma_i,$$

---

[20]Possible extensions include the one-sided or two-sided discontinuous cases.

and substituting back into (3.82) yields

$$\sum_{i=0}^{s-1} \gamma_i \int_0^1 P_j(\tau)P_i(\tau)\mathrm{d}\tau = \int_0^1 P_j(\tau)f(\sigma(\tau h))\mathrm{d}\tau,$$

or, exploiting the orthogonality property of Legendre polynomials,

$$\gamma_j = \int_0^1 P_j(\tau)f(\sigma(\tau h))\mathrm{d}\tau, \qquad j = 0,\ldots,s-1.$$

that is the projected problem (3.75) defined over the shifted and normalized Legendre basis. As is easily argued, the use of a interpolatory quadrature formula to approximate the integral appearing in (3.81) defines a HBVM(k, s) method in the form (3.76) written along the Legendre basis.

Bibliographical notes

HBVMs have been presented in [36, 37, 38, 40, 48] (see also [118, 119, 120]); their analysis based on the local Fourier expansion is taken from [45] (see also [41]); their isospectrality has been studied in [44].

We refer the reader to [105, Chapter XI] for a thorough analysis based on KAM perturbation theory, showing the importance of the property of symmetry of the methods.

The AVF method has been presented in [149] (see also [132]); its discretization has been studied in [68]; related B-series methods are in [69].

The EPCM methods can be found in [103] (see also [108] for the study of their conjugate-symplecticity).

Time finite element methods are described in [166]; earlier references on the approach are [16, 17, 18, 116, 117, 165].

Chapter 4

Implementing the methods and numerical illustrations

We here describe, in some detail, the actual implementation of HBVMs. In fact, in order to exploit all the potentialities of the methods, they must be carefully and efficiently coded.

4.1 Fixed-point iterations

As was seen in Section 3.2, a HBVM(k, s) method is a k-stage implicit Runge-Kutta method of order $2s$. In view of the result presented in Theorem 3.4 on page 91, k is usually chosen (arbitrarily) larger than s, thus it is not convenient to implement a HBVM the way it is done for a classical Runge-Kutta method, since the discrete problem generated by the computation of the stages has (block) dimension k. We shall instead compute the s coefficients $\hat{\gamma}_j$ of the corresponding polynomial approximation (see (3.22)–(3.24) on page 89)

$$u(ch) = y_0 + h \sum_{j=0}^{s-1} \int_0^c P_j(x)\mathrm{d}x\, \hat{\gamma}_j, \qquad c \in [0, 1].$$

In more details, in order to solve the local problem (see (3.1)–(3.4))

$$\dot{y}(ch) = f(y(ch)), \qquad c \in [0, 1], \qquad y(0) = y_0 \in \mathbb{R}^{2m}, \tag{4.1}$$

by using the HBVM(k, s) method based at the k Gauss-Legendre abscissae $c_1 < \cdots < c_k$, we can introduce the following (block) vectors:

$$\hat{\gamma} = \begin{pmatrix} \hat{\gamma}_0 \\ \vdots \\ \hat{\gamma}_{s-1} \end{pmatrix}, \qquad Y = \begin{pmatrix} Y_1 \\ \vdots \\ Y_k \end{pmatrix}, \qquad f(Y) = \begin{pmatrix} f(Y_1) \\ \vdots \\ f(Y_k) \end{pmatrix}, \tag{4.2}$$

containing the coefficients of the polynomial $u(ch)$, the stages of the Runge-Kutta formulation, and the evaluation of f at such stages. Consequently, by considering the matrices $\mathcal{I}_s$, $\mathcal{P}_s$, and Ω defined at (3.33)–(3.35), the following relations hold true:

$$Y = \mathbf{1} \otimes y_0 + h \mathcal{I}_s \otimes I_{2m}\,\hat{\gamma}, \qquad \hat{\gamma} = \mathcal{P}_s^\top \Omega \otimes I_{2m}\,f(Y). \qquad (4.3)$$

They express, in vector form, the first set of k equations in (3.31) and the s equations in (3.23), respectively: the former defining the stages of the Runge-Kutta method, the latter defining the coefficients of the polynomial $u(ch)$.[1] Substituting the first equation in (4.3) into the second yields the following set of s (block) equations:

$$\hat{\gamma} = \mathcal{P}_s^\top \Omega \otimes I_{2m}\,f\left(\mathbf{1} \otimes y_0 + h\mathcal{I}_s \otimes I_{2m}\,\hat{\gamma}\right), \qquad (4.4)$$

which defines the final shape of the discrete problem to be solved at each integration step.

The discrete problem (4.4) can be solved in several ways, the simplest one being the following *fixed-point iteration:*[2]

$$\hat{\gamma}^{k+1} \ := \ \mathcal{P}_s^\top \Omega \otimes I_{2m}\,f\left(\mathbf{1} \otimes y_0 + h\mathcal{I}_s \otimes I_{2m}\,\hat{\gamma}^k\right), \quad k = 0, 1, \ldots, \quad (4.5)$$

which is iterated until full accuracy is obtained. In fact, as is easily argued, solving only approximately (4.5) implies the method to lose, in general, its conservation properties. The new approximation is then obtained according to (3.24), namely:

$$y_1 := y_0 + h\hat{\gamma}_0. \qquad (4.6)$$

Concerning the convergence of the iteration (4.5) the following result is easily established (see also Theorem 3.1).

Theorem 4.1. *Let $f(y)$ be continuously differentiable, so that there exists $L > 0$ such that $\|f'(y)\| \leqslant L$ in a neighborhood of y_0.[3] Then, the iteration (4.5) converges, provided that $hL \leqslant \frac{5}{4}$.*

Proof. Let us set $e^k := \hat{\gamma}^k - \hat{\gamma}$ the error at the iteration k. Subtracting (4.4) from (4.5), one obtains:

$$e^{k+1} \equiv \hat{\gamma}^{k+1} - \hat{\gamma}$$

$$= \mathcal{P}_s^\top \Omega \otimes I_{2m}\left[f\left(\mathbf{1} \otimes y_0 + h\mathcal{I}_s \otimes I_{2m}\,\hat{\gamma}^k\right) - f\left(\mathbf{1} \otimes y_0 + h\mathcal{I}_s \otimes I_{2m}\,\hat{\gamma}\right)\right]$$

$$= h\mathcal{P}_s^\top \Omega \mathcal{I}_s \otimes I_{2m} \int_0^1 f'\left(\mathbf{1} \otimes y_0 + h\mathcal{I}_s \otimes I_{2m}(\hat{\gamma} + \tau(\hat{\gamma}^k - \hat{\gamma}))\right)(\hat{\gamma}^k - \hat{\gamma})\mathrm{d}\tau$$

$$= h\,X_s \otimes I_{2m}\left[\int_0^1 f'\left(\mathbf{1} \otimes y_0 + h\mathcal{I}_s \otimes I_{2m}(\hat{\gamma} + \tau e^k)\right)\mathrm{d}\tau\right]e^k.$$

---

[1] As is clear, hereafter $\mathbf{1} \in \mathbb{R}^k$.

[2] The initial vector can be chosen, e.g., as $\hat{\gamma}^0 = \mathbf{0}$.

[3] L is "essentially" the local Lipschitz constant of f, in a neighborhood of y_0. Moreover, $\|\cdot\|$ denotes the 1, 2, or ∞-norm.

The claim then follows since $\left\| f' \left(\mathbf{1} + h \mathcal{I}_s \otimes I_{2m} (\hat{\gamma} + \tau e^k) \right) \right\| \leq L$ by hypothesis, and, moreover, $\| X_s \| \leq (\sqrt{3}+1)/(2\sqrt{3}) < \frac{4}{5}$, as is easily seen from (3.37).[4] Consequently, for $e^k \neq \mathbf{0}$ one has

$$\| e^{k+1} \| < \left(\frac{4}{5} h L \right) \| e^k \| \leq \| e^k \|,$$

provided that $hL \leq \frac{5}{4}$. □

4.1.1 Estimating the local errors

After solving (4.4), one obtains the new approximation (4.6) to $y(h)$, solution of (4.1), which has a $O(h^{2s+1})$ local error (since the HBVM(k,s) method has order $2s$). In order to obtain an estimate of the local error for y_1, we assume now $k > s$, and consider the approximation, of order $2s + 2$, provided by the HBVM$(k, s + 1)$ method: it is obtained by solving the corresponding discrete problem (compare with (4.4))

$$\bar{\gamma} = \mathcal{P}_{s+1}^{\top} \Omega \otimes I_{2m} \, f \left(\mathbf{1} \otimes y_0 + h \mathcal{I}_{s+1} \otimes I_{2m} \, \bar{\gamma} \right), \qquad (4.7)$$

where $\bar{\gamma} = \left(\, \bar{\gamma}_0^{\top}, \quad \ldots, \quad \bar{\gamma}_{s-1}^{\top}, \quad \bar{\gamma}_s^{\top} \, \right)^{\top}$ has now block-size $s + 1$, instead of s. The new approximation is then obtained as $\bar{y}_1 := y_0 + h \bar{\gamma}_0$. Consequently, the difference $h(\bar{\gamma}_0 - \hat{\gamma}_0)$ provides us with an estimate of the local error for y_1. Equation (4.7) may be solved by means of a suitable iterative procedure (e.g., (4.5)), starting from the initial approximation $\bar{\gamma}^0 = \left(\, \hat{\gamma}^{\top}, \quad 0^{\top} \, \right)^{\top}$, which is expected to be very close to the solution. Very few iterations are usually enough to obtain a sufficiently accurate approximation and, consequently, a good error estimate.

4.1.2 The case of special second-order problems

It is possible to devise a more efficient procedure, when dealing with special second order problems, namely problems in the form

$$\ddot{q} = \nabla U(q), \qquad t \geq 0, \qquad q(0) = q_0, \qquad \dot{q}(0) = p_0, \qquad q_0, p_0 \in \mathbb{R}^m, \quad (4.8)$$

which frequently occur in the applications: as matter of fact, most of the problems described in Chapter 2 are of this type. Problem (4.8) is indeed Hamiltonian, as is clearly seen when casting it into first order form,

$$\dot{q} = p, \qquad \dot{p} = \nabla U(q), \qquad t \geq 0, \qquad q(0) = q_0, \qquad p(0) = p_0, \qquad (4.9)$$

corresponding to the following Hamiltonian function:

$$H(q, p) = \frac{1}{2} p^{\top} p - U(q). \qquad (4.10)$$

---

[4]For the 2-norm, this follows from the fact that $\| X_s \|_2 \leq \sqrt{\| X_s \|_1 \| X_s \|_\infty}$.

In such a case, the stage vector Y in (4.2) can be partitioned into two (block) vectors Q and P containing the stages for q and p, respectively. According to the Runge-Kutta formulation of a HBVM(k, s) method (see Theorem 3.6 on page 95), such vectors are then formally obtained as

$$Q = \mathbf{1} \otimes q_0 + h \mathcal{I}_s \mathcal{P}_s^\top \Omega \otimes I_m\, P, \tag{4.11}$$

$$P = \mathbf{1} \otimes p_0 + h \mathcal{I}_s \mathcal{P}_s^\top \Omega \otimes I_m\, \nabla U(Q), \tag{4.12}$$

where, by using a notation similar to (4.2), $\nabla U(Q)$ contains the evaluation of ∇U at the corresponding (block) entries of Q. The new approximations are then given by:

$$q_1 = q_0 + h e_1^\top \mathcal{P}_s^\top \Omega \otimes I_m\, P, \qquad p_1 = p_0 + h e_1^\top \mathcal{P}_s^\top \Omega \otimes I_m\, \nabla U(Q), \tag{4.13}$$

with $e_1 \in \mathbb{R}^s$ the first vector of the canonical basis. However, by substituting (4.12) into (4.11), we obtain a problem of halved dimension, since it involves only the vector Q:

$$Q = \mathbf{1} \otimes q_0 + h \mathcal{I}_s \mathcal{P}_s^\top \Omega \otimes I_m \left[\mathbf{1} \otimes p_0 + h \mathcal{I}_s \mathcal{P}_s^\top \Omega \otimes I_m\, \nabla U(Q) \right].$$

By considering that, from Lemma 3.6 on page 96:

- $\mathcal{I}_s \mathcal{P}_s^\top \Omega \mathbf{1} = \mathcal{I}_s e_1 \equiv c,$ with c the vector of the abscissae of the Runge-Kutta method,

- $\mathcal{I}_s \mathcal{P}_s^\top \Omega \mathcal{I}_s \mathcal{P}_s^\top \Omega = \mathcal{I}_s X_s \mathcal{P}_s^\top \Omega,$

one obtains that

$$Q = \mathbf{1} \otimes q_0 + h c \otimes p_0 + h^2 \mathcal{I}_s X_s \mathcal{P}_s^\top \Omega \otimes I_m\, \nabla U(Q). \tag{4.14}$$

By setting

$$\hat{\gamma} := \mathcal{P}_s^\top \Omega \otimes I_m\, \nabla U(Q) \equiv \left(\hat{\gamma}_0^\top, \quad \ldots, \quad \hat{\gamma}_{s-1}^\top \right)^\top, \tag{4.15}$$

we finally arrive at the following discrete problem of (block) size s:

$$\hat{\gamma} = \mathcal{P}_s^\top \Omega \otimes I_m\, \nabla U \left(\mathbf{1} \otimes q_0 + h c \otimes p_0 + h^2 \mathcal{I}_s X_s \otimes I_m\, \hat{\gamma} \right), \tag{4.16}$$

whose block dimension is half that of the corresponding formulation (4.4).[5] Provided that the stepsize h is suitably small, this problem can be solved by using a fixed-point iteration analogous to (4.5):

$$\hat{\gamma}^{k+1} := \mathcal{P}_s^\top \Omega \otimes I_m\, \nabla U \left(\mathbf{1} \otimes q_0 + h c \otimes p_0 + h^2 \mathcal{I}_s X_s \otimes I_m\, \hat{\gamma}^k \right),$$
$$k = 0, 1, \ldots, \tag{4.17}$$

---

[5] Apart from the dimensionality, the two problems are equivalent to each other, as is seen by taking into account the form of the problem (4.9).

which is iterated until full accuracy is obtained. The new approximations, q_1 and p_1, are then obtained, by virtue of (4.11)–(4.13) and (4.15)-(4.16), as:

$$\begin{aligned} q_1 &= q_0 + h e_1^\top \mathcal{P}_s^\top \Omega \otimes I_m \left(1 \otimes p_0 + h \mathcal{I}_s \otimes I_m \hat{\gamma}\right) \\ &= q_0 + h p_0 + h^2 e_1^\top X_s \otimes I_m \hat{\gamma}, \\ p_1 &= p_0 + h e_1^\top \otimes I_m \hat{\gamma}. \end{aligned}$$

That is, by considering that $\frac{1}{2}$ and $-\xi_1 = -1/(2\sqrt{3})$ are the first two entries on the first row of matrix X_s defined at (3.37) (see page 96):

$$q_1 = q_0 + h p_0 + \frac{h^2}{2}\left(\hat{\gamma}_0 - \frac{1}{\sqrt{3}}\hat{\gamma}_1\right), \qquad p_1 = p_0 + h\hat{\gamma}_0. \tag{4.18}$$

It is not difficult to check that the new approximations in (4.18) are such that $q_1 \equiv w(h)$ and $p_1 \equiv v(h)$, where

$$\begin{aligned} v(ch) &= p_0 + h \sum_{j=0}^{s-1} \int_0^c P_j(x)\mathrm{d}x\, \hat{\gamma}_j, \\ w(ch) &= q_0 + h \sum_{j=0}^{s-1} \int_0^c P_j(x)\mathrm{d}x\, \phi_j(v), \qquad c \in [0,1], \end{aligned}$$

are polynomials of degree s, (locally) approximating $p(ch)$ and $q(ch)$, respectively. In the expression of the polynomial $w(ch)$, the vector coefficients $\phi_j(v)$ are defined as (see (3.37)):

$$\phi_j(v) := \int_0^1 P_j(\tau)v(\tau h)\mathrm{d}\tau = \begin{cases} p_0 + h(\frac{1}{2}\hat{\gamma}_0 - \xi_1 \hat{\gamma}_1), & \text{if} \quad j = 0, \\ h(\xi_j \hat{\gamma}_{j-1} - \xi_{j+1}\hat{\gamma}_{j+1}), & \text{if} \quad 1 \leqslant j < s-1, \\ h\xi_{s-1}\hat{\gamma}_{s-2}, & \text{if} \quad j = s-1, \end{cases}$$

due to property **P9** of the Legendre polynomials listed in Section A.1. In fact, by setting the (block) vector

$$\phi = \left(\phi_0(v)^\top, \quad \ldots, \quad \phi_{s-1}(v)^\top \right)^\top,$$

one has, by virtue of (4.12), (4.15), and Lemma 3.6:

$$\begin{aligned} \phi &= \mathcal{P}_s^\top \Omega \otimes I_m\, P = \mathcal{P}_s^\top \Omega \otimes I_m \left(1 \otimes p_0 + h \mathcal{I}_s \otimes I_m \hat{\gamma}\right) \\ &= e_1 \otimes p_0 + h X_s \otimes I_m \hat{\gamma}. \end{aligned}$$

The above procedure can be generalized, in a quite straightforward way, to the case where the Hamiltonian takes the form

$$H(q,p) = \frac{1}{2}p^\top M p - U(q), \tag{4.19}$$

instead of (4.10), where M is a symmetric (and, usually, also positive definite) matrix.[6] In such a case, (4.9) becomes

$$\dot{q} = Mp, \qquad \dot{p} = \nabla U(q), \qquad t \geq 0, \qquad q(0) = q_0, \qquad p(0) = p_0, \qquad (4.20)$$

and, accordingly, (4.14) becomes:

$$Q = \mathbf{1} \otimes q_0 + h c \otimes M p_0 + h^2 \mathcal{I}_s X_s \mathcal{P}_s^\top \Omega \otimes M \, \nabla U(Q).$$

Consequently, by setting $\hat{\gamma}$ as in (4.15), one arrives at the dicrete problem

$$\hat{\gamma} = \mathcal{P}_s^\top \Omega \otimes I_m \, \nabla U \left(\mathbf{1} \otimes q_0 + h c \otimes M p_0 + h^2 \mathcal{I}_s X_s \otimes M \, \hat{\gamma} \right). \qquad (4.21)$$

This latter can be solved through the following fixed-point iteration,

$$
\begin{aligned}
\hat{\gamma}^{k+1} &:= \mathcal{P}_s^\top \Omega \otimes I_m \, \nabla U \left(\mathbf{1} \otimes q_0 + h c \otimes M p_0 + h^2 \mathcal{I}_s X_s \otimes M \, \hat{\gamma}^k \right), \\
& \quad k = 0, \ldots,
\end{aligned}
\qquad (4.22)
$$

which generalizes (4.17) to the present case. After that, the new approximations are given by (compare with (4.18))

$$q_1 = q_0 + h M p_0 + \frac{h^2}{2} M \left(\hat{\gamma}_0 - \frac{1}{\sqrt{3}} \hat{\gamma}_1 \right), \qquad p_1 = p_0 + h \hat{\gamma}_0.$$

where, as usual, $\hat{\gamma}_0$ and $\hat{\gamma}_1$ are the first two (block) entries of the vector $\hat{\gamma}$ defined at (4.15).

4.2 Newton-like iterations

If one wants to improve on stepsize limitations that may occur when dealing with *stiff oscillatory Hamiltonian problems*, for which the constant L defined in Theorem 4.1 may be very large, one has to resort to a Netwon-type method for solving the discrete problem (4.4). A popular choice is to consider the *simplified Newton iteration*. In more details, by setting

$$F(\hat{\gamma}) := \hat{\gamma} - \mathcal{P}_s^\top \Omega \otimes I_{2m} \, f \left(\mathbf{1} \otimes y_0 + h \mathcal{I}_s \otimes I_{2m} \hat{\gamma} \right), \qquad (4.23)$$

and $f'(y_0)$ the Jacobian matrix of f evaluated at y_0, one iterates:[7]

$$
\begin{aligned}
\text{solve}: & \quad \left[I - h \mathcal{P}_s^\top \Omega \mathcal{I}_s \otimes f'(y_0) \right] \Delta^k = -F(\hat{\gamma}^k), \qquad (4.24) \\
\text{set}: & \quad \hat{\gamma}^{k+1} = \hat{\gamma}^k + \Delta^k, \qquad k = 0, 1, \ldots.
\end{aligned}
$$

---

[6]As is clear, (4.10) can be obtained from (4.19) as the special case $M = I_m$.

[7]As is clear, now I denotes the identity matrix of dimension $2sm$.

By considering that, as was seen above, $\mathcal{P}_s^\top \Omega \mathcal{I}_s = X_s$, the iteration (4.24) becomes

$$\text{solve}: \quad \left[I - hX_s \otimes f'(y_0)\right]\Delta^k = -F(\hat{\gamma}^k), \tag{4.25}$$

$$\text{set}: \quad \hat{\gamma}^{k+1} = \hat{\gamma}^k + \Delta^k, \quad k = 0, 1, \dots.$$

Consequently, the bulk of the iteration consists in the solution of the linear systems in the first equation of (4.25), whose coefficient matrix is, however, *independent of* k. A straight implementation of the iteration (4.25) would then require the factorization of the coefficient matrix $[I - hX_s \otimes f'(y_0)]$, having dimension $2sm \times 2sm$, where $2m$ is the dimension of the continuous problem. Nevertheless, there are convenient alternatives to this approach: in the following, we shall describe one of them in detail.[8]

The starting point is that the first equation in (4.25) can be written in the following two equivalent forms (we neglect the superscript k of the iteration, for sake of brevity):

$$\left[I - hX_s \otimes f'(y_0)\right]\Delta \;=\; -F(\hat{\gamma}) \;=:\; \boldsymbol{\eta}, \tag{4.26}$$

$$\left[\rho_s X_s^{-1} \otimes I_{2m} - h\rho_s I_s \otimes f'(y_0)\right]\Delta \;=\; \rho_s X_s^{-1} \otimes I_{2m}\, \boldsymbol{\eta} \;=:\; \boldsymbol{\eta}_1, \tag{4.27}$$

where ρ_s is a positive parameter, to be specified later. Consequently:

(blend1) for $h \approx 0$, (4.26) becomes $\left[I + O(h)\right]\Delta = \boldsymbol{\eta}$;

(blend2) for $h \gg 1$, (4.27) becomes $h\left[O(h^{-1}) - \rho_s I_s \otimes f'(y_0)\right]\Delta = \boldsymbol{\eta}_1$.

Next, let us define the *weighting function*

$$\Theta = I_s \otimes \Psi^{-1}, \quad \Psi := I_{2m} - h\rho_s f'(y_0) \in \mathbb{R}^{2m \times 2m}. \tag{4.28}$$

We note that, if λ is an eigenvalue of $f'(y_0)$, then matrix Ψ^{-1}, restricted to the invariant subspace relative to λ, is approximately given by:

(blend3) the identity matrix, if $h\lambda \approx 0$;

(blend4) the zero matrix, if $h|\lambda| \gg 1$.

We can now define the *blended equation* corresponding to (4.25), which is given by:

$$\left[\Theta\left[I - hX_s \otimes f'(y_0)\right] + (I - \Theta)\left[\rho_s X_s^{-1} \otimes I_{2m} - h\rho_s I_s \otimes f'(y_0)\right]\right]\Delta$$
$$= \Theta\boldsymbol{\eta} + (I - \Theta)\boldsymbol{\eta}_1. \tag{4.29}$$

It is obtained as the *blending* of the two equivalent formulations (4.26)-(4.27) with weights Θ and $(I-\Theta)$, respectively. By taking into account the properties

---

[8]References to additional approaches can be found in the bibliographical notes.

(blend1)–(blend4) listed above, one obtains that the coefficient matrix of the linear system (4.29) can be approximated as

$$\Theta \left[I - hX_s \otimes f'(y_0) \right] + (I - \Theta) \left[\rho_s X_s^{-1} \otimes I_{2m} - h\rho_s I_s \otimes f'(y_0) \right]$$
$$\approx \ I - h\rho_s I_s \otimes f'(y_0) \ \equiv \ I_s \otimes \Psi,$$

where matrix Ψ is that defined at (4.28). Consequently, one obtains the following approximation to the solution of (4.29):

$$\Delta \approx \Theta \left[\Theta \eta + (I - \Theta) \eta_1 \right] = \Theta \left[\eta_1 + \Theta \left(\eta - \eta_1 \right) \right].$$

In conclusion, we can approximate (4.25), with the following *blended iteration*:

$$
\begin{aligned}
&\text{factor matrix}: \quad \Psi := I_{2m} - h\rho_s f'(y_0) \\
&\text{for } k = 0, 1, \ldots : \\
&\qquad \text{compute}: \quad \eta^k := -F(\hat{\gamma}^k), \quad \eta_1^k := \rho_s X_s^{-1} \otimes I_{2m}\, \eta^k \\
&\qquad \text{solve}: \quad I_s \otimes \Psi\, u^k = \eta^k - \eta_1^k \\
&\qquad \text{solve}: \quad I_s \otimes \Psi\, \Delta^k = \eta_1^k + u^k \qquad\qquad (4.30)\\
&\qquad \text{set}: \quad \hat{\gamma}^{k+1} = \hat{\gamma}^k + \Delta^k
\end{aligned}
$$

which only requires the factorization of matrix $\Psi \in \mathbb{R}^{2m \times 2m}$. Concerning the positive parameter ρ_s, its choice is done in order to optimize the convergence properties of the blended iteration (4.30). For this purpose, due to the particular structure of the spectrum of matrix X_s, it is possible to prove that the best choice consists in requiring that matrix $\rho_s X_s^{-1}$ has spectral radius 1.[9] Consequently,

$$\rho_s = \min_{\lambda \in \sigma(X_s)} |\lambda|. \qquad\qquad (4.31)$$

In Table 4.1 we list some of such parameters for increasing values of s. It is worth noticing that these values apply to a HBVM(k, s) method, having fixed s, *independently* of the chosen value of k.

4.2.1 The case of special second-order problems

In the case of Hamiltonian problems in the form (4.20), the simplified Newton iteration applied for solving (4.21) reads, by using the same notation of Section 4.1.2:

$$\text{solve}: \quad \left[I - h^2 X_s^2 \otimes M \nabla^2 U(q_0) \right] \Delta^k = -F(\hat{\gamma}^k), \qquad (4.32)$$

$$\text{set}: \quad \hat{\gamma}^{k+1} = \hat{\gamma}^k + \Delta^k, \qquad k = 0, 1, \ldots.$$

with

$$F(\hat{\gamma}) := \hat{\gamma} - \mathcal{P}_s^\top \Omega \otimes I_m\, \nabla U \left(1 \otimes q_0 + hc \otimes M p_0 + h^2 \mathcal{I}_s X_s \otimes M\, \hat{\gamma} \right).$$

---

[9]See [49] for details.

TABLE 4.1: Optimal value for the parameter (4.31) of the *blended iteration* (4.30).

s	ρ_s
1	5.0000e-01
2	2.8868e-01
3	1.9673e-01
4	1.4752e-01
5	1.1734e-01
6	9.7103e-02
7	8.2651e-02
8	7.1846e-02
9	6.3479e-02
10	5.6817e-02

Considering that (4.32) is formally the same as (4.25) via the subsitutions

$$f'(y_0) \to M\nabla^2 U(q_0), \qquad h \to h^2, \qquad X_s \to X_s^2, \tag{4.33}$$

one can repeat the same arguments as above. Thus, introducing the further substitution

$$\rho_s \to \rho_s^2$$

besides (4.33), one eventually arrives at the following *blended iteration*:

$$
\begin{aligned}
\text{factor matrix}: \quad & \Psi := I_m - h^2 \rho_s^2 M\nabla^2 U(q_0) \\
\text{for } k = 0, 1, \ldots: \quad & \\
\text{compute}: \quad & \eta^k := -F(\hat{\gamma}^k), \quad \eta_1^k := \rho_s^2 X_s^{-2} \otimes I_m \, \eta^k \\
\text{solve}: \quad & I_s \otimes \Psi \, u^k = \eta^k - \eta_1^k \\
\text{solve}: \quad & I_s \otimes \Psi \, \Delta^k = \eta_1^k + u^k \\
\text{set}: \quad & \hat{\gamma}^{k+1} = \hat{\gamma}^k + \Delta^k
\end{aligned}
\tag{4.34}
$$

which is the analogue of (4.30) for the second-order problem (4.20). The best choice for the parameter ρ_s turns out to be still given by (4.31). Again, one has only to factor a matrix Ψ, having the same size as that of the continuous problem (i.e., the dimension of q_0).

4.2.2 Matrix form of the discrete problems

For the actual coding of the iterations that solve the nonlinear system $F(\hat{\gamma}) = \mathbf{0}$, with F given by (4.23), it could be more convenient to recast it in *matrix form*, which yields a more compact formulation of the discrete problem with respect to the original vector form. With reference to (4.1)-

(4.2), by setting

$$\Gamma \;\; = \;\; \left(\; \hat{\gamma}_0, \;\; \ldots, \;\; \hat{\gamma}_{s-1} \; \right) \in \mathbb{R}^{2m \times s},$$
$$\bar{Y} \;\; = \;\; \left(\; Y_1, \;\; \ldots, \;\; Y_k \; \right) \in \mathbb{R}^{2m \times k}, \qquad (4.35)$$
$$f(\bar{Y}) \;\; = \;\; \left(\; f(Y_1), \;\; \ldots, \;\; f(Y_k) \; \right) \in \mathbb{R}^{2m \times k},$$

we see that the first equation in (4.3) is equivalent to

$$\bar{Y} = y_0 \mathbf{1}^\top + h \Gamma \mathcal{I}_s^\top.$$

Analogously, one also derives that (4.4) is equivalent to

$$\Gamma = f(y_0 \mathbf{1}^\top + h \Gamma \mathcal{I}_s^\top) \Omega \mathcal{P}_s.$$

Consequently, the fixed-point iteration (4.5) can be recast as

$$\Gamma^{k+1} := f(y_0 \mathbf{1}^\top + h \Gamma^k \mathcal{I}_s^\top) \Omega \mathcal{P}_s, \qquad k = 0, 1, \ldots.$$

Similarly, the *blended iteration* (4.30) becomes:[10]

$$\Sigma^k \;\; := \;\; -\Gamma^k + f(y_0 \mathbf{1}^\top + h \Gamma^k \mathcal{I}_s^\top) \Omega \mathcal{P}_s,$$
$$\Sigma_1^k \;\; := \;\; \Sigma^k (\rho_s X_s^{-\top}),$$
$$\Gamma^{k+1} \;\; := \;\; \Gamma^k + \Psi^{-1} \left[\Sigma_1^k + \Psi^{-1} (\Sigma^k - \Sigma_1^k) \right], \qquad k = 0, 1, \ldots.$$

In case of special second order problems (4.20), corresponding matrix formulations of (4.21), (4.22), and (4.34) are similarly obtained in a quite straightforward way.

We conclude this section by observing that it could be sometimes convenient to define the matrices Γ, $\bar{Y}$, and $f(\bar{Y})$ as the transposes of (4.35), i.e.,

$$\Gamma = \begin{pmatrix} \hat{\gamma}_0^\top \\ \vdots \\ \hat{\gamma}_{s-1}^\top \end{pmatrix}, \quad \bar{Y} = \begin{pmatrix} Y_1^\top \\ \vdots \\ Y_k^\top \end{pmatrix}, \quad f(\bar{Y}) = \begin{pmatrix} f(Y_1)^\top \\ \vdots \\ f(Y_k)^\top \end{pmatrix} \in \mathbb{R}^{k \times 2m},$$

in order to improve locality of data, depending on the used programming language. In such a case, the previous matrix equations are replaced by the corresponding transpose ones.

4.3   Recovering round-off and iteration errors

The implementation of implicit energy-conserving methods clearly requires that the used nonlinear iteration has to be executed to full accuracy. Conversely, the conservation properties of the method may not be valid, according

---

[10]For sake of brevity, let us formally use Ψ^{-1}.

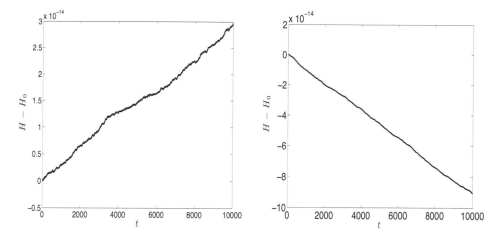

FIGURE 4.1: Error in the numerical Hamiltonian when solving the Hénon-Heiles problem (4.36)-(4.37) by using the HBVM(3,2) method with stepsize $h = 0.25$. Left-plot: discrete problem solved via blended iteration. Right-plot: discrete poblem solved via fixed-point iteration. As is clear, in both cases a numerical drift occurs.

to the analysis made in Chapter 3 (see, e.g., Theorem 3.4 on page 91). Nonetheless, both round-off and iteration errors (RIEs) are inevitably introduced at each integration step: in fact, a zero tolerance for the stopping criterion of the nonlinear iteration *cannot* be considered and, obviously, on a computer a finite precision arithmetic is used. As a result, the accumulation of such errors may cause a (artificial) numerical drift of the Hamiltonian, even in the case where exact conservation is expected. The following simple example clearly elucidates this problem.

Let us consider the Hénon-Heiles problem described in Section 2.3 (see page 55) which is defined by the polynomial Hamiltonian (see (2.7)–(2.8))

$$H(q_1, q_2, p_1, p_2) = \frac{1}{2}\left(p_1^2 + p_2^2\right) + U(q_1, q_2), \tag{4.36}$$

$$U(q_1, q_2) = \frac{1}{2}\left(q_1^2 + q_2^2\right) + q_2\left(q_1^2 - \frac{q_2^2}{3}\right),$$

with the initial condition

$$\left(\begin{array}{cccc} q_1(0), & q_2(0), & p_1(0), & p_2(0) \end{array} \right)^\top := \left(\begin{array}{cccc} 0.1, & -0.5, & 0, & 0 \end{array} \right)^\top. \tag{4.37}$$

Since the Hamiltonian (4.36) is a polynomial of degree 3, according to Theorem 3.4, HBVM(k, s) methods are exactly energy-conserving for all $k \geq \frac{3}{2}s$. Nonetheless, because of the use of finite-precision arithmetic and of an iterative procedure for solving the generated discrete problems (e.g., those studied in the initial part of the chapter), RIEs are introduced at each integration step that, though very small, eventually accumulate, thus resulting in a numerical drift, as stated above. This fact is confirmed by the plots in Figure 4.1, where the HBVM(3,2) method is used for integrating the problem over the interval $[0, 10^4]$, by using a stepsize $h = 0.25$. In particular, the left-plot is obtained by using a blended iteration for solving the discrete problem, whereas the right-plot is obtained by using a fixed-point iteration. As is clear, in both cases one obtains a numerical drift, though of different shape. However, since the error at each step is very small, we can easily recover this drawback, by using the following straightforward remedy.

When approximating $y_1 \approx y(h)$ starting from y_0, so that $H(y_1) = H(y_0)$, the nonlinear iteration is stopped at the first iteration $k + 1$ satisfying

$$\|(y_1^{k+1} - y_1^k)./(1 + |y_0|)\|_2 \leqslant \varepsilon\sqrt{2m}, \tag{4.38}$$

where ε is a (quite small) tolerance, $./$ stands for the component-wise division between vectors, y_1^k is the k-th iterate approximating y_1, $|y_0|$ is the vector with the absolute values of the corresponding entries of y_0, and $2m$ is the length of the given vectors. In particular, ε is of the order of the used machine precision. Clearly, the left-hand side of the inequality in (4.38) is an estimate of the error

$$\|(y_1 - y_1^k)./(1 + |y_0|)\|_2.$$

Consequently, having got the new approximation y_1, after stopping the iteration according to (4.38), we see that if $H(y_1) \neq H(y_0)$ but

$$|H(y_1) - H(y_0)| \lesssim \varepsilon\sqrt{2m} \max\{1, \|y_1\|_2\} \|\nabla H(y_1)\|_2 \tag{4.39}$$

holds true, this means that the error in the numerical Hamiltonian is not due to the quadrature but to the (unavoidable) round-off and iteration errors. It is then reasonable to slightly "tune" y_1 to obtain a better approximation, say $\bar{y}_1$, which reduces the difference $|H(y_1) - H(y_0)|$. For this purpose, we look for $\bar{y}_1$ in the form

$$\bar{y}_1 := y_1 + \alpha\nabla H(y_1), \tag{4.40}$$

where α is obtained by means of one Newton step applied to the equation $g(\alpha) := H(y_1 + \alpha\nabla H(y_1)) - H(y_0) = 0$, starting from 0. Then, one has:

$$\alpha = -\frac{H(y_1) - H(y_0)}{\|\nabla H(y_1)\|_2^2}. \tag{4.41}$$

It is worth noticing that this remedy is quite inexpensive. In fact, for canonical Hamiltonian problems, one obtains that (see (3.2)-(3.3) on page 82)

$$\nabla H(y_1) = J^\top f(y_1),$$

where f is the right-hand side of the differential equation. Nevertheless, it turns out to be quite effective. As an example, in Figure 4.2, we plot the error in the numerical Hamiltonian obtained by using again the HBVM(3,2)

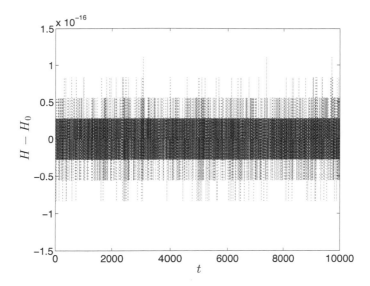

FIGURE 4.2: Error in the numerical Hamiltonian when solving the Hénon-Heiles problem (4.36)-(4.37) through the HBVM(3,2) method with stepsize $h = 0.25$, by using the correction (4.40)-(4.41), provided that (4.39) holds true.

method with stepsize $h = 0.25$ for solving problem (4.36)-(4.37) on the interval $[0, 10^4]$. In such a case, however, the correction (4.40)-(4.41) is performed at the end of each integration step, provided that (4.39) holds true. As is clear, no numerical drift now occurs.

For sake of completeness, we end this section by mentioning that a kind of "mean correction" can be also used by considering, in (4.40)-(4.41), $J^\top \hat\gamma_0$ in place of $\nabla H(y_1)$. In fact,

$$J^\top \hat\gamma_0 \approx \int_0^1 \nabla H(u(\tau h)) \mathrm{d}\tau$$

which can be interpreted as a "mean gradient" of H in the interval $[0, h]$.

4.4 Numerical illustrations

We here report a few numerical tests involving the use of HBVMs. For this purpose, the problems described in Chapter 2 will be considered as test problems.[11] One of the goals of these experiments is to emphasize the role

---

[11]The code used for the numerical tests is described in Section A.2.

that the energy-conservation property could play in reproducing the correct behavior of the solution in the phase space. To this end, we have deliberately considered very special situations, such as the dynamics in a neighborhood of a separatrix, where the benefits deriving from the use of an energy-conserving method might emerge more clearly. Therefore, we would like to warn the reader that the results presented in this chapter depict only a narrow range of the vastly rich spectrum of behaviors exhibited by a Hamiltonian system, and by no means are these examples intended to push other paramount features of geometric integrators available in the literature into the background.

Kepler problem

As was shown in Section 2.5, this problem is defined by the Hamiltonian (see (2.18) on page 60)

$$H(q_1, q_2, p_1, p_2) = \frac{1}{2} \left(p_1^2 + p_2^2 \right) - \frac{1}{\sqrt{q_1^2 + q_2^2}}, \qquad (4.42)$$

and the trajectory starting at

$$\left(\; q_1(0), \quad q_2(0), \quad p_1(0), \quad p_2(0) \; \right) := \left(\; 1 - e, \quad 0, \quad 0, \quad \sqrt{\frac{1+e}{1-e}} \; \right) \qquad (4.43)$$

for $e \in [0, 1)$, is periodic of period 2π and describes an ellipse of eccentricity e and major semi-axis 1 in the (q_1, q_2)-plane. Denoting by $r(t)$ the distance of the point mass from the origin, according to (4.43), at time $t = 0$ the point mass is located at the periapsis (the closer point of the ellipse to the origin) and $r(0) = r_{\min} = 1 - e$, while after a half-period it approaches the apoapsis (the farthest point of the ellipse from the origin) at distance of $r_{\max} = 1 + e$ from the origin (see Figure 2.6 on page 64).

It is well-known that the closer is e to 1, the harder is the numerical integration of the problem. In fact, the ratio between the maximum and minimum distances is

$$\frac{r_{\max}}{r_{\min}} = \frac{1+e}{1-e}$$

and becomes unbounded as $e \to 1^-$. Consequently, the force acting on the point mass, which behaves as r^{-2}, undergoes huge variations in magnitude during the motion when $e \simeq 1$ and the same conclusion applies to its derivatives. In such a case, conserving the energy may improve the accuracy of the computed solution. In order to make this point clear, let us consider the following two 6-th order methods for solving the problem: HBVM(3,3) (i.e., the 3-stage Gauss-Legendre method) and HBVM(12,3). We use the stepsize $h = \pi/200$, and integrate problem (4.42)-(4.43) over 100 periods, when the eccentricity $e \in \{0.5, 0.6, 0.7, 0.8, 0.9\}$. The results are summarized in Figure 4.3:

- in the left-plot is the Hamiltonian error along the numerical solution. In more detail, for the i-th period, we plot the maximum value of

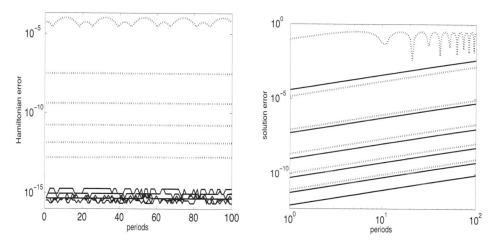

FIGURE 4.3: Numerical solution of the Kepler problem with periodic solution of eccentricity $e \in \{0.4, 0.5, 0.6, 0.7, 0.8, 0.9\}$ by using HBVM(3,3) and HBVM(12,3) with stepsize $h = \pi/200$. Left-plot: Hamiltonian error over 100 periods for HBVM(3,3) (dotted lines) and HBVM(12,3) (solid lines). Right-plot: solution error over 100 periods for HBVM(3,3) (dotted lines) and HBVM(12,3) (solid lines). In both plots, larger values of e are associated to larger errors, except for the Hamiltonian error produced by HBVM(12,3), which is always of the order of the machine precision.

$|H_n - H_0|$, for $n \in \{(i-1)400 + 1, \ldots, i400\}$. The dotted lines are relative to HBVM(3,3), whereas the solid ones concern HBVM(12,3). Since the solid lines are all of the order of round-off errors, we see that the aptitude of this latter method in conserving the energy is not influenced by the value of the eccentricity and the numerical Hamiltonian is practically conserved. On the other hand, the dotted lines shows an increasing behavior, starting from $e = 0.4$ (the bottom one) up to $e = 0.9$ (the top one), even though no drift occurs;

- in the right-plot is the solution error for HBVM(3,3) (dotted lines) and HBVM(12,3) (solid lines), measured at each period. For both methods, the error curves at the bottom of the picture correspond to the lowest value of the eccentricity ($e = 0.4$), while those at the top come out from the highest value of the eccentricity ($e = 0.9$). All the curves in between bring the intermediate values of e in increasing order with e. Consequently, we see that the larger e, the larger the error in the solution. It is worth noticing that the error growth is linear and, moreover, the error provided by the energy-conserving HBVM(12,3) method is approximately 2 orders of magnitude smaller than that provided by HBVM(3,3), for the same value of the eccentricity.

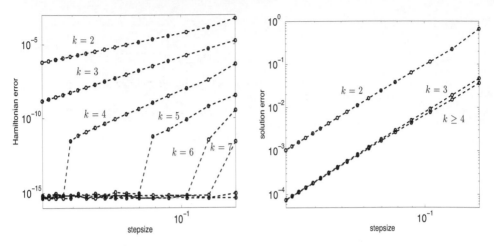

FIGURE 4.4: Numerical solution of the Kepler problem with periodic solution of eccentricity $e = 0.6$ by using HBVM(k,2), $k = 2, \ldots, 10$, with stepsizes $h_i = 10^{-1}\pi/(i+1)$, $i = 1, \ldots, 9$, on 10 periods. Left-plot: Hamiltonian error. Right-plot: solution error.

Looking at these results, we can argue that the energy-conservation property can prove useful in improving the overall approximation properties of the method, especially when the problem becomes more difficult to solve.

Next, let us consider the numerical solution of problem (4.42)-(4.43) over 10 periods, when the eccentricity is $e = 0.6$. We now use the HBVM(k, s) methods, $k = 2, \ldots, 10$, with stepsize $h_i = 10^{-1}\pi/(i+1)$, $i = 1, \ldots, 9$. The obtained results are summarized in the two plots in Figure 4.4:

- the left-plot shows the Hamiltonian error versus the stepsize used: as one may see, the Hamiltonian error decreases with the prescribed order $2k$, until full machine accuracy is obtained. In particular, for $k \geqslant 8$, the Hamiltonian error turns out to be negligible;

- the right-plot shows the solution error versus the stepsize used: all methods are clearly 4-th order, even though the error decreases with k and, for $k \geqslant 4$, the corresponding errors are almost identical.

Cassini ovals

Let us now consider the problem of the *Cassini ovals*, described in Section 2.2, which, for $a^2 = 5$ (see (2.5) on page 54), is defined by the following Hamiltonian function:

$$H(q,p) = \left(q^2 + p^2\right)^2 - 10\left(q^2 - p^2\right). \tag{4.44}$$

The following initial condition,

$$q(0) = 0, \qquad p(0) = 10^{-6}, \tag{4.45}$$

yields a Cassini oval with a single loop which is characterized, according to (2.6), by the ratio

$$r := b/a \approx 1 + 10^{-13},$$

i.e., very close to the lemniscate, obtained for $r = 1$. Consequently, the corresponding solution is periodic, and its period is

$$T \approx 3.131990057003955.$$

We solve the problem by using a stepsize $h = 10^{-2}T$, and with the following fourth-order methods: HBVM(4,2), which is energy-conserving, since the Hamiltonian has degree 4, and HBVM(2,2) (i.e., the symplectic 2-stage Gauss method). The obtained results, summarized by the plots in Figure 4.5, are described below:

- the energy-conserving HBVM(4,2) is able to stay on the correct level-set of the Hamiltonian, as is shown in the two upper-plots in Figure 4.5. In particular, the phase portrait close to the origin is quite well reproduced, as is shown in the enlarged view on the right;

- the solution provided by the symplectic 2-stage Gauss method fails to lie on the correct level-set of the Hamiltonian, as is shown in the two middle plots of Figure 4.5. Since we are integrating the system in a close neighborhood of the separatrix, even small displacements from the correct level set may cause the orbit to drastically change its global behavior;

- the bottom-left picture of Figure 4.5 displays the Hamiltonian error of the two computed solutions, and confirms that it is negligible for HBVM(4,2), whereas it is not for HBVM(2,2). As a result, a correct periodicity pattern is reproduced by HBVM(4,2), as is shown in the plot on the right, while the numerical solution computed by HBVM(2,2) behaves erratically, since close to the origin it may randomly jump on nearby level sets of the Hamiltonian function, as described in the item above.

We then conclude that, also in this case, energy conservation may play a significant role in providing a reliable numerical simulation in those regions of the phase space where the phase portrait sensibly depends on the value of the energy.

Fermi-Pasta-Ulam problem

This problem, described in Section 2.7 (see page 76), is interesting since it provides a concrete example of *stiff oscillatory problem*. As a consequence,

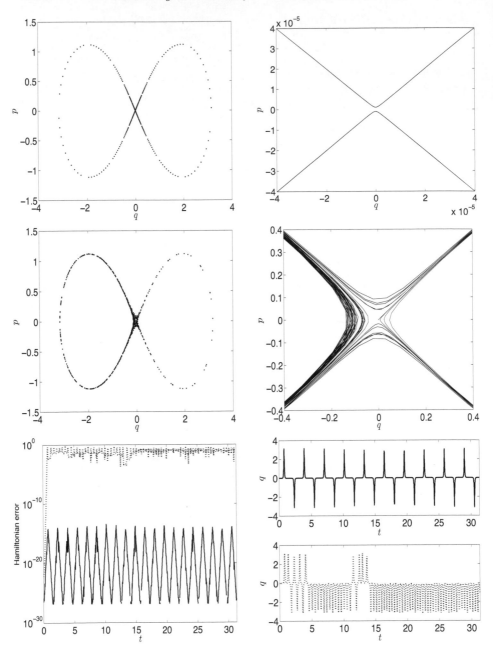

FIGURE 4.5: Numerical solution of the Cassini problem (4.44)-(4.45) by using a stepsize $h = 10^{-2}T$. Upper plots: phase-space solution computed by HBVM(4,2); intermediate plots: phase-space solution computed by HBVM(2,2); lower plots: left - Hamiltonian error for HBVM(4,2) (solid line) and HBVM(2,2) (dotted line); right - q component computed by HBVM(4,2) (up) and HBVM(2,2) (down).

the (quite inexpensive) fixed-point iteration would require very small stepsizes, resulting completely ineffective. The same conclusion applies to explicit methods, e.g., the Störmer-Verlet method or composition methods based on it. Conversely, as we will see, the use of a suitable HBVM(k, s) method, implemented via a Newton-like iteration (e.g., the blended iteration described in Section 4.2) would allow for much coarser stepsizes.

First of all, since the Hamiltonian (see (2.51)) is a polynomial of degree 4,

$$H(q, p) = \frac{1}{2} \sum_{i=1}^{m} (p_{2i-1}^2 + p_{2i}^2) + \frac{1}{4} \sum_{i=1}^{m} \omega_i^2 (q_{2i} - q_{2i-1})^2 + \sum_{i=0}^{m} (q_{2i+1} - q_{2i})^4, \quad (4.46)$$

with $q, p \in \mathbb{R}^{2m}$ and $q_0 = q_{2m+1} = 0$, all HBVM($2s, s$) methods turn out to be energy-conserving and of order $2s$. In particular, we shall consider the sixth-order HBVM(6,3) method for numerically solving the problem with

$$m = 7, \qquad \omega_i = \omega_{8-i} = 10, \ i = 1, 2, 3, \quad \text{and} \quad \omega_4 = 10^4, \qquad (4.47)$$

the initial condition and the end time being

$$p_i(0) = 0, \qquad q_i(0) = \frac{i - 1}{2(2m - 1)}, \qquad i = 1, \ldots, 2m, \qquad T = 10. \quad (4.48)$$

With these initial conditions, most of the components (24 out of 28) turn out to be slowly oscillating, whereas components q_7, q_8, p_7, and p_8 undertake high oscillations. As said above, the use of a fixed-point iteration turns out to be quite inefficient, in this case. As matter of fact, the iteration works only when using stepsizes not larger than $5 \cdot 10^{-4}$. The sixth-order composition method based on Störmer-Verlet, in turn, requires the use of a stepsize not larger than 10^{-4}, thus confirming that explicit methods are generally inefficient for *stiff* problems. On the other hand, by using the blended iteration for solving the discrete problems generated by HBVM(6,3), any stepsize is allowed in the practice: one could even use a stepsize as large as $h = 1$ and still get a convergent iteration. However, in order to obtain a reasonably accurate solution for the low-frequency components, a stepsize $h = 0.25$ is at least recommended.

In Figure 4.6 is the plot of the two high-frequency components q_8 and p_8 (left), and of the two low-frequency components q_9 and p_9 (right), by using HBVM(6,3) with stepsizes $h = 10^{-4}$ (solid line) and $h = 0.25$ (circles). As one may see, the low-frequency components are quite well-matched, as well as the amplitude of the component q_8.[12] In both cases, the relative error in the numerical Hamiltonian is smaller than $2 \cdot 10^{-15}$. Moreover, the blended implementation of HBVM(6,3) with stepsize $h = 0.25$ requires only 875 iterations for covering the whole integration interval.

---

[12] A stepsize $h = 0.1$ is required for matching the amplitude of p_8, too.

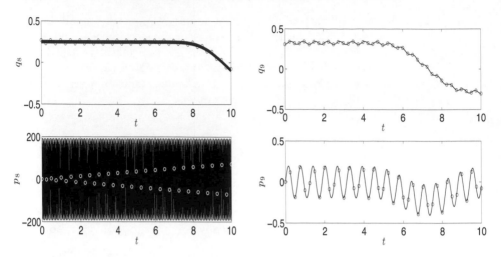

FIGURE 4.6: Numerical solution of the Fermi-Pasta-Ulam problem (4.46)-(4.48) by using HBVM(6,3) with stepsizes $h = 10^{-4}$ (solid line) and $h = 0.25$ (circles). Left-plot: high-frequency components q_8 and p_8. Right-plot: low-frequency components q_9 and p_9.

Hénon-Heiles problem

Let us consider again the Hénon-Heiles problem described in Section 2.3 and defined by the Hamiltonian recalled at (4.36). We know that

$$\mathcal{U} := \left\{ (q_1, q_2)^\top \in \mathbb{R}^2 : U(q_1, q_2) \leqslant \frac{1}{6} \right\}$$

defines the equilateral triangle with vertices:

$$Q_1 = (0, 1)^\top, \qquad Q_2 = \left(-\frac{\sqrt{3}}{2}, -\frac{1}{2} \right)^\top, \qquad Q_3 = \left(\frac{\sqrt{3}}{2}, -\frac{1}{2} \right)^\top.$$

Consequently, any orbit starting at a point $\left(q_1(0), q_2(0), p_1(0), p_2(0) \right)^\top$ such that $H(q_1(0), q_2(0), p_1(0), p_2(0)) < \frac{1}{6}$, will remain inside $\mathcal{U}$ forever. To highlight the benefits that could emerge from using an energy-conserving method, we consider the initial point

$$\left(q_1(0), q_2(0), p_1(0), p_2(0) \right)^\top := \left(0, \ -\tfrac{1}{8}, \ \tfrac{99}{10^4}, \ -\sqrt{\left(\tfrac{3}{4} \right)^4 - 10^{-4}} \right)^\top, \tag{4.49}$$

which satisfies

$$H(q_1(0), q_2(0), p_1(0), p_2(0)) \approx \frac{1}{6} - 9.95 \cdot 10^{-7},$$

so that a small perturbation of the energy caused by the use of any non-conservative method could lead to an erroneous simulation of the system. We solve the problem defined by (4.36) and (4.49) by using the following fourth-order methods with stepsize $h = 1$ for $2 \cdot 10^4$ integration steps:

- the symplectic 2-stage Gauss method (i.e., HBVM(2,2)), and

- the HBVM(3,2) method, which is energy-conserving due to the fact that the Hamiltonian is a polynomial of degree 3.

Figure 4.7 summarizes the obtained results:

- the pictures on the left concern the HBVM(2,2) method: in the upper one is the Hamiltonian error; in the bottom one is the computed solution in the q_1-q_2 plane. After approximately 11600 integration steps, the numerical Hamiltonian blows-up causing the solution to escape the triangle and to eventually diverge towards infinity. This is confirmed by the lower plot, where the first 4 points exiting the triangle $\mathcal{U}$ are highlighted by a circle;

- the pictures on the right concern the HBVM(3,2) method: in the upper one is the Hamiltonian error; in the bottom one is the computed solution in the q_1-q_2 plane. In this case, the Hamiltonian error is negligible and the numerical solution completely lies inside $\mathcal{U}$, filling it densely.

Nonlinear pendulum

This problem, described in Section 2.1, is defined by the Hamiltonian (see (2.2) on page 51)

$$H(q, p) = \frac{1}{2}p^2 - \cos(q), \tag{4.50}$$

and its solutions lie on the the level curves $H = const \geq -1$. In particular, one has: a stable equilibrium point for $H = -1$; periodic oscillations around the stable equilibrium point for $|H| < 1$; rotations for $H > 1$. Fixing $q(0) = 0$, the value separating the last two regimes is $p(0) = 2$, since $H(0, 2) = 1$. Consequently, by considering the initial condition

$$q(0) = 0, \qquad p(0) = 2 - \varepsilon, \qquad 0 < \varepsilon \approx 0, \tag{4.51}$$

one has $H \approx 1 - \varepsilon$ and, then, the trajectories should be periodic. In particular, let us fix $\varepsilon = 10^{-5}$, and consider the numerical solutions of such problem generated by the following fourth-order methods with a stepsize $h = 0.5$:

- the symplectic 2-stage Gauss method (i.e., HBVM(2,2)), and

- the HBVM(8,2) method, which is *practically* energy-conserving for the problem at hand.

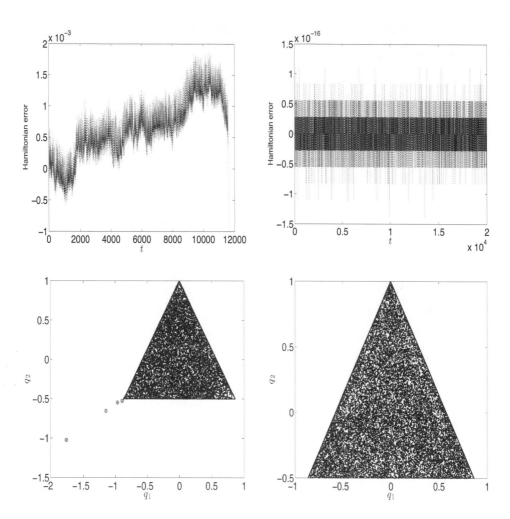

FIGURE 4.7: Numerical solution of the Hénon-Heiles problem (4.36)-(4.49) by using the HBVM(2,2) and HBVM(3,2) methods with stepsize $h = 1$. Left-plots: HBVM(2,2). Right-plots: HBVM(3,2).

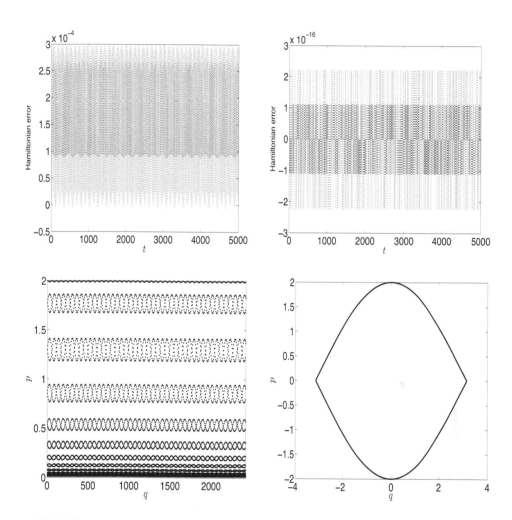

FIGURE 4.8: Numerical solutions of the pendulum problem (4.50)-(4.51) for $\varepsilon = 10^{-5}$, computed by the HBVM(2,2) and HBVM(8,2) methods with stepsize $h = 0.5$. Left-plots: HBVM(2,2). Right-plots: HBVM(8,2).

Figure 4.8 summarizes the results:

- the pictures on the left concern HBVM(2,2): in the upper one is the Hamiltonian error; in the bottom one is the computed solution. As one may see, the Hamiltonian error, though not exhibiting a drift, is large enough to modify the qualitative behavior of the computed solution, which is now of rotation type;

- the plots on the right concern HBVM(8,2): in the upper one is the Hamiltonian error, which is negligible; in the bottom one is the computed solution, which is qualitatively correct, i.e., periodic.

Planar circular restricted three-body problem

This problem has been described in Section 2.6.1 (see page 70), and is interesting, because its numerical solution may be very challenging, when a quasi-collision occurs since, in such an event, the problem is close to be singular. The problem is defined by the Hamiltonian (see (2.42)),

$$H(q, p) = \tag{4.52}$$

$$p_1 q_2 - p_2 q_1 + \frac{1}{2}\left(p_1^2 + p_2^2\right) - \frac{1 - \mu}{\sqrt{(q_1 + \mu)^2 + q_2^2}} - \frac{\mu}{\sqrt{(q_1 - 1 + \mu)^2 + q_2^2}},$$

where $m_2 := \mu \in (0, 1)$ and $m_1 := 1 - \mu$ are the (normalized) masses of the two primaries, which are located at the points $(-\mu, 0)$ and $(1 - \mu, 0)$ in the q_1-q_2 plane. We consider the following parameter and initial point,

$$\mu = 0.3, \qquad \begin{pmatrix} q_1(0) \\ q_2(0) \\ p_1(0) \\ p_2(0) \end{pmatrix} := \begin{pmatrix} 6.955415851757456e - 01 \\ 6.456133089565784e - 03 \\ -7.663758431829502e + 00 \\ 4.716929963135788e + 00 \end{pmatrix}. \tag{4.53}$$

The corresponding trajectory, which is shown in the upper plot of Figure 4.9, starts close to one of the primaries, located at $(0.7, 0)$, and has to be contained in the region of the q_1-q_2 plane defined by (see (2.45))

$$\Omega(q_1, q_2) \geqslant -H(q(0), p(0)) \approx 3.55642,$$

depicted in white in the figure, where

$$\Omega(q_1, q_2) := \frac{1}{2}(q_1^2 + q_2^2) + \frac{(1 - \mu)}{\sqrt{(q_1 + \mu)^2 + q_2^2}} + \frac{\mu}{\sqrt{(q_1 - 1 + \mu)^2 + q_2^2}}$$

is the effective potential (2.43). We then compute the numerical trajectories, in the interval [0,5], by means of the following methods:

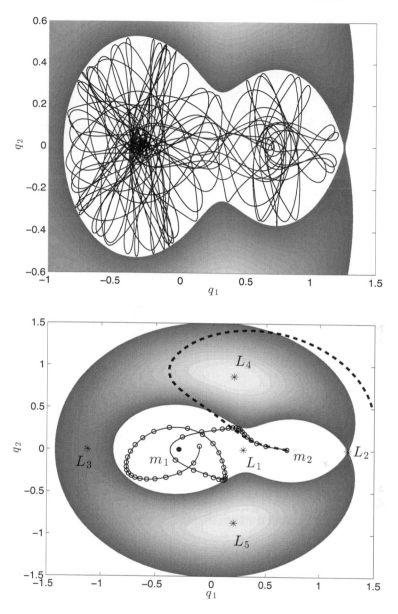

FIGURE 4.9: Solution of the planar restricted three-body problem (4.52)-(4.53). Upper plot: trajectory starting at (4.53) in the interval [0,100] (solid line), along with the invariant region containing it. Lower plot: invariant region along with the numerically computed trajectories starting at (4.53), in the interval [0,5], by using: HBVM(1,1) with stepsize $h = 5 \cdot 10^{-4}$ (dashed line); HBVM(8,1) with stepsize $h = 5 \cdot 10^{-4}$ (circles); HBVM(9,3) with variable stepsize and tolerance $tol = 10^{-10}$ (solid line). In both pictures, the grayscale filled area identifies the forbidden region. In the lower plot, L_1–L_5 denote the Lagrangian points.

TABLE 4.2: initial conditions for problem (4.54)-(4.55).

atom	1	2	3	4	5	6	7
position	0.00	0.02	0.34	0.36	-0.02	-0.35	-0.31
	0.00	0.39	0.17	-0.21	-0.40	-0.16	0.21
velocity	-30	50	-70	90	80	-40	-80
	-20	-90	-60	40	90	100	-60

- the symplectic implicit mid-point method (i.e., HBVM(1,1)), with a constant stepsize $h = 5 \cdot 10^{-4}$;

- the (practically) energy-conserving method HBVM(8,1), with a constant stepsize $h = 5 \cdot 10^{-4}$;

- the HBVM(9,3) method with variable stepsize, tolerance $tol = 10^{-10}$, and initial stepsize $h_0 = 5 \cdot 10^{-4}$. This solution, computed with high precision, is assumed as the reference trajectory.

The last two methods provide a practical conservation of energy, along the numerical trajectory, whereas the first method fails to conserve the energy (error $\simeq 0.4$).[13] The lower plot in Figure 4.9 summarizes the obtained results: the energy-conserving methods provide a similar trajectory, which is contained in the prescribed region, whereas the symplectic method produces a wrong numerical trajectory, which exits the admissible region (i.e., the Hill region).

It must be stressed that this problem becomes very difficult to be handled numerically when quasi-collisions occur since, in such a case, the magnitude of the gravitational forces may become arbitrarily large causing the problem to get close to a singularity.[14] Consequently, a constant stepsize is not recommended. As matter of fact, the trajectory depicted in the upper picture of Figure 4.9 has been computed by using HBVM(9,3) with a variable stepsize strategy, based on the estimate of the local error sketched in Section 4.1.1. In particular, at each integration step, the stepsize is modulated in order to match, by means of a standard extrapolation procedure, the requirement

$$\|e./(1 + |y|)\| = tol,$$

where e is the error estimate, y is the obtained approximation, $./$ is the componentwise division, and tol is the prescribed tolerance. To compute the trajectory in the upper plot of Figure 4.9 a tolerance $tol = 10^{-10}$ has been used, along with an initial stepsize $h_0 = 5 \cdot 10^{-4}$. In so doing, only $N = 6079$ steps are needed to cover the interval $[0,100]$, by selecting stepsizes in the range $[1.35 \cdot 10^{-7}, 0.15]$.

---

[13]We recall that for this problem there are no first integrals other than the Hamiltonian function itself.

[14]We do not consider regularizations of singularities here.

Molecular dynamics

This problem has been described in its general form in Section 2.8 on page 79. Here we consider a particular instance, consisting of a frozen argon crystal with N atoms evolving in a plane, with positions q_i and momenta p_i. Since the mass of each atom is the same, as well as all interactions laws among them, one has, with reference to (2.53):

$$m_i = m, \qquad \varepsilon_{ij} = \varepsilon, \qquad \sigma_{ij} = \sigma, \qquad \forall i,j = 1,\dots,N.$$

Consequently, the Hamiltonian simplifies to

$$H(\boldsymbol{q},\boldsymbol{p}) = \frac{1}{2m}\sum_{i=1}^{N}\|p_i\|_2^2 + 4\varepsilon\sum_{i=1}^{N-1}\sum_{j=i+1}^{N} V(\|q_i - q_j\|_2), \qquad (4.54)$$

where $p_i = m\dot{q}_i$, and

$$V(r) = \left(\frac{\sigma}{r}\right)^6\left[\left(\frac{\sigma}{r}\right)^6 - 1\right].$$

The parameters in (4.54) are chosen as follows:

$$
\begin{aligned}
N &= 7,\\
m &= 66.34\cdot 10^{-27} \ \ [\text{kg}],\\
\sigma &= 0.341 \ \ [\text{nm}],\\
\varepsilon &= 119.8\,\kappa_B \ \ [\text{J}],
\end{aligned}
\qquad (4.55)
$$

where

$$\kappa_B = 1.380658\cdot 10^{-23} \ \ [\text{J/K}]$$

is the Boltzmann constant. By choosing the initial conditions according to Table 4.2 the Hamiltonian has a value

$$H_0 \approx -1260.2\,\kappa_B \ [\text{J}].$$

In the left-plot of Figure 4.10 the initial condition is depicted, whereas on the right is the plot of the relative error in the numerical Hamiltonian, obtained by simulating the problem by using the HBVM(8,2) method with a constant stepsize $h = 10^{-4}$ [nsec]. As is clear from this latter figure, the error in the Hamiltonian is of the order of round-off errors.

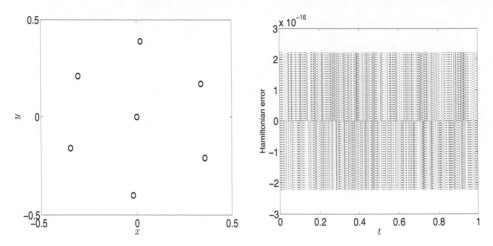

FIGURE 4.10: Left-plot: initial condition for the problem (4.54)-(4.55) shown in Table 4.2 (positions only). Right-plot: relative error in the numerical Hamiltonian when approximating the problem by using HBVM(8,2) with a stepsize $h = 10^{-4}$ [sec].

Bibliographical notes

The blended implementation of HBVMs is inherited from that of block implicit methods [49], used in the computational codes BiM and BiMD [50, 51], for ODE-IVPs and linearly implicit DAEs. The extension to HBVMs has been done in [43]. Related approaches have been considered in [26, 27, 33].

The efficient implementation of Runge-Kutta methods for ODE-IVPs has been studied by several authors (see, e.g., [5, 61, 112, 113]). A thorough linear analysis of convergence for splitting iterations, generalizing that in [112, 113], is given in [53].

The recovery of round-off errors in the implementation of HBVMs is sketched in [35]. The same problem has been also considered by other authors, in the more general framework of geometric integration (see, e.g., [77, 85, 160]).

The numerics of linear matrix equations can be found in [158]. The problem of the frozen argon crystal is taken from [105, pages 18–20].

Additional material concerning the above numerical tests can be retrieved at the homepage of the book, as is specified in Section A.2. In particular, for each problem, a related *graphics interchange format (gif)* file is provided, along with Matlab functions and software; we refer the reader to Section A.2 for more details.

Chapter 5

Hamiltonian Partial Differential Equations

In this chapter we deal with the application of HBVMs for solving Hamiltonian PDEs, which are characterized, as for the ordinary differential case, by the presence of conservation laws. These invariants are replaced by corresponding discrete analogues which are required to be precisely conserved by the discrete solution.

5.1 The semilinear wave equation

In this chapter we discuss energy-conservation issues for *Hamiltonian Partial Differential Equations*. In particular, we consider the semilinear wave equation, though the approach can be extended to different kinds of Hamiltonian PDEs, as sketched in Section 5.5 for the nonlinear Schrödinger equation. For sake of simplicity, we shall consider the following 1D case,

$$
\begin{aligned}
u_{tt}(x,t) &= u_{xx}(x,t) - f'(u(x,t)), & (x,t) \in (0,1) \times [0,\infty), \\
u(x,0) &= \psi_0(x), & \\
u_t(x,0) &= \psi_1(x), & x \in (0,1),
\end{aligned} \tag{5.1}
$$

coupled with appropriate boundary conditions. As usual, subscripts denote partial derivatives. In (5.1), the functions f, ψ_0 and ψ_1 are supposed to be sufficiently smooth and such that they define a regular solution $u(x,t)$ (f' denotes the derivative of f). The problem is completed by assigning suitable boundary conditions which, at first, we shall assume to be periodic:

$$
u(0,t) = u(1,t), \qquad t \geqslant 0. \tag{5.2}
$$

In such a case, we will also assume that ψ_0, ψ_1 and f are such that the resulting solution also satisfies [1]

$$
u_x(0,t) = u_x(1,t), \qquad t \geqslant 0. \tag{5.3}
$$

---

[1] In particular, in case of periodic boundary conditions, we will assume u to be at least C^4 in space, as a periodic function on $[0,1]$.

Later on, we shall also consider the case of Dirichlet boundary conditions,

$$u(0,t) = \varphi_0(t), \qquad u(1,t) = \varphi_1(t), \qquad t \geqslant 0, \tag{5.4}$$

and Neumann boundary conditions,

$$u_x(0,t) = \phi_0(t), \qquad u_x(1,t) = \phi_1(t), \qquad t \geqslant 0, \tag{5.5}$$

with $\varphi_0(t)$, $\varphi_1(t)$, $\phi_0(t)$, and $\phi_1(t)$ suitably regular. We set

$$v = u_t, \tag{5.6}$$

and define the *Hamiltonian functional*

$$\mathcal{H}[u,v](t) = \int_0^1 \left[\frac{1}{2}v^2(x,t) + \frac{1}{2}u_x^2(x,t) + f(u(x,t)) \right] dx =: \int_0^1 E(x,t)\, dx. \tag{5.7}$$

As is well known, we can rewrite (5.1) as the infinite-dimensional Hamiltonian system [2]

$$z_t = J \frac{\delta \mathcal{H}}{\delta z}, \tag{5.8}$$

where

$$J = \begin{pmatrix} 0 & 1 \\ -1 & 0 \end{pmatrix}, \qquad z = \begin{pmatrix} u \\ v \end{pmatrix}, \tag{5.9}$$

and

$$\frac{\delta \mathcal{H}}{\delta z} = \left(\frac{\delta \mathcal{H}}{\delta u}, \frac{\delta \mathcal{H}}{\delta v} \right)^{\mathsf{T}} \tag{5.10}$$

is the functional derivative of $\mathcal{H}$. This latter is defined as follows: given a generic functional in the form

$$\mathcal{L}[q] = \int_a^b L\left(x, q(x), q'(x), q''(x), \ldots, q^{(n)}(x), \ldots \right) dx,$$

with $q^{(n)}(x)$ the n-th derivative of $q(x)$, the functional derivative of $\mathcal{L}[q]$ is defined as

$$\frac{\delta \mathcal{L}}{\delta q} = \sum_{n \geqslant 0} (-1)^n \frac{d^n}{dx^n} \frac{\partial L}{\partial q^{(n)}},$$

Consequently, in the particular case where $L = L(x, q, q')$, one obtains:

$$\frac{\delta \mathcal{L}}{\delta q} = \frac{\partial L}{\partial q} - \left(\frac{d}{dx} \frac{\partial L}{\partial q'} \right). \tag{5.11}$$

Exploiting (5.11), one then easily verifies that (5.8)–(5.10) are equivalent to (5.1):

$$z_t = \begin{pmatrix} u_t \\ v_t \end{pmatrix} = J \frac{\delta \mathcal{H}}{\delta z} = \begin{pmatrix} \frac{\delta \mathcal{H}}{\delta v} \\ -\frac{\delta \mathcal{H}}{\delta u} \end{pmatrix} = \begin{pmatrix} v \\ u_{xx} - f'(u) \end{pmatrix},$$

---

[2] For sake of brevity, we shall sometimes omit the arguments of the functions u and v.

or

$$u_t(x,t) = v(x,t), \qquad (x,t) \in (0,1) \times [0,\infty),$$
$$v_t(x,t) = u_{xx}(x,t) - f'(u(x,t)), \qquad (5.12)$$

that is, the first-order formulation of the first equation in (5.1).

By considering that the time derivative of the integrand function $E(x,t)$ defined at (5.7) satisfies (see (5.12))

$$
\begin{aligned}
E_t(x,t) &= v(x,t)v_t(x,t) + u_x(x,t)u_{xt}(x,t) + f'(u(x,t))u_t(x,t) \\
&= v(x,t)(u_{xx}(x,t) - f'(u(x,t))) + u_x(x,t)v_x(x,t) + f'(u(x,t))v(x,t) \\
&= v(x,t)u_{xx}(x,t) + u_x(x,t)v_x(x,t) = (u_x(x,t)v(x,t))_x,
\end{aligned}
$$

one obtains the *conservation law*

$$E_t(x,t) + F_x(x,t) = 0, \qquad \text{with} \qquad F(x,t) = -u_x(x,t)v(x,t). \qquad (5.13)$$

This relation allows us to derive the expression for the time derivative of the Hamiltonian:[3]

$$
\begin{aligned}
\dot{\mathcal{H}}[z](t) &= \int_0^1 E_t(x,t)\mathrm{d}x = -\int_0^1 F_x(x,t)\mathrm{d}x = F(0,t) - F(1,t) \\
&\equiv [u_x(1,t)v(1,t) - u_x(0,t)v(0,t)]. \qquad (5.14)
\end{aligned}
$$

In particular, by taking into account (5.6), we see that the derivative of the Hamiltonian vanishes when the prescribed boundary conditions for problem (5.1) are:

i) periodic boundary conditions, i.e. (5.2) (and, then, (5.3)) $\Rightarrow$ $u_x(1,t)v(1,t) = u_x(0,t)v(0,t)$;

ii) constant Dirichlet boundary conditions, i.e. (5.4) with $\dot{\varphi}_0(t) = \dot{\varphi}_1(t) = 0$ $\Rightarrow$ $v(1,t) = v(0,t) \equiv 0$;

iii) homogeneous Neumann boundary conditions, i.e. (5.5) with $\phi_0(t) = \phi_1(t) = 0$ $\Rightarrow$ $u_x(1,t) = u_x(0,t) \equiv 0$.

In such cases, the Hamiltonian (5.7) is a conserved quantity. Moreover, considering that the following identity holds true,

$$u_x^2(x,t) = (u(x,t)u_x(x,t))_x - u(x,t)u_{xx}(x,t),$$

we can rewrite the Hamiltonian (5.7) in the equivalent form:

$$
\begin{aligned}
\mathcal{H}[z](t) &= \int_0^1 E(x,t)\,\mathrm{d}x = \\
&= \int_0^1 \left[\frac{1}{2}v^2(x,t) - \frac{1}{2}u(x,t)u_{xx}(x,t) + f(u(x,t))\right]\mathrm{d}x \\
&\quad + \frac{1}{2}[u(1,t)u_x(1,t) - u(0,t)u_x(0,t)], \qquad (5.15)
\end{aligned}
$$

---

[3]Hereafter, we shall denote the arguments of $\mathcal{H}$ either by u, v, or z, according to (5.9).

which simplifies to

$$\mathcal{H}[z](t) = \int_0^1 \left[\frac{1}{2}v^2(x,t) - \frac{1}{2}u(x,t)u_{xx}(x,t) + f(u(x,t)) \right] dx, \qquad (5.16)$$

in the cases **i)** and **iii)** listed above. In case **ii)**, (5.16) holds true only when homogeneous Dirichlet boundary conditions, i.e. $\varphi_0(t) = \varphi_1(t) = 0$ for all $t \geqslant 0$, are prescribed for the problem.

We are now going to study, in more detail, the two cases where the boundary conditions prescribed for the probem are:

- periodic (i.e., given by (5.2)) or

- of Dirichlet or Neumann type (i.e., given by (5.4) or (5.5), respectively).

5.2 Periodic boundary conditions

A popular way of solving time dependent PDEs, such as (5.1) (or equivalently (5.12)), consists in:

- discretize, at first, the spatial variable;

- then solve the resulting system of ordinary differential equations.

In accord with this strategy, let us define the set of abscissae

$$x_i = i\Delta x, \quad i = 0, \dots, N, \qquad \Delta x = 1/N, \qquad (5.17)$$

and the following second-order approximations to the second derivative in (5.1) evaluated at the abscissae (5.17):

$$u_{xx}(x_i,t) \approx \frac{u(x_{i+1},t) - 2u(x_i,t) + u(x_{i-1},t)}{\Delta x^2}. \qquad (5.18)$$

Consequently, we replace (5.1) with the following set of differential equations,

$$\ddot{u}_i(t) = \frac{u_{i+1}(t) - 2u_i(t) + u_{i-1}(t)}{\Delta x^2} - f'(u_i(t)), \quad i = 0, \dots, N-1, \quad t \geqslant 0, \qquad (5.19)$$

where $u_i(t)$ is the resulting approximation to $u(x_i,t)$. In a similar way, according to (5.6), one obtains

$$v_i(t) = \dot{u}_i(t), \qquad i = 0, \dots, N-1, \qquad t \geqslant 0, \qquad (5.20)$$

as the corresponding approximation to $v(x_i,t)$. Evidently, in (5.19) we have taken into account that, because of the periodic boundary conditions, $u_N(t) =$

$u_0(t)$, and $u_{-1}(t) = u_{N-1}(t)$. Moreover, according to (5.1), the initial conditions for (5.19)-(5.20) are given by

$$u_i(0) = \psi_0(x_i), \qquad v_i(0) = \psi_1(x_i), \qquad i = 0, \ldots, N-1.$$

Defining the vectors

$$\boldsymbol{x} = \begin{pmatrix} x_0 \\ \vdots \\ x_{N-1} \end{pmatrix}, \quad \boldsymbol{q} = \begin{pmatrix} u_0 \\ \vdots \\ u_{N-1} \end{pmatrix}, \quad \boldsymbol{p} = \begin{pmatrix} v_0 \\ \vdots \\ v_{N-1} \end{pmatrix} \in \mathbb{R}^N,$$

and the matrix

$$T_N = \begin{bmatrix} 2 & -1 & & & -1 \\ -1 & \ddots & \ddots & & \\ & \ddots & \ddots & \ddots & \\ & & \ddots & \ddots & -1 \\ -1 & & & -1 & 2 \end{bmatrix} \in \mathbb{R}^{N \times N}, \tag{5.21}$$

the previous problem can be cast in vector form as

$$\begin{aligned} \dot{\boldsymbol{q}} &= \boldsymbol{p}, \\ \dot{\boldsymbol{p}} &= -\frac{1}{\Delta x^2} T_N \boldsymbol{q} - f'(\boldsymbol{q}), \qquad t \geq 0, \end{aligned} \tag{5.22}$$

with the initial conditions

$$\boldsymbol{q}(0) = \psi_0(\boldsymbol{x}), \qquad \boldsymbol{p}(0) = \psi_1(\boldsymbol{x}),$$

(with an obvious meaning for $f'(\boldsymbol{q})$, $\psi_0(\boldsymbol{x})$, and $\psi_1(\boldsymbol{x})$). By setting

$$J_N = \frac{1}{\Delta x} \begin{pmatrix} & I_N \\ -I_N & \end{pmatrix} \qquad \text{and} \qquad \boldsymbol{y} = \begin{pmatrix} \boldsymbol{q} \\ \boldsymbol{p} \end{pmatrix},$$

the following result holds clearly true.

Theorem 5.1. *Problem (5.22) can be written in Hamiltonian form as*

$$\dot{\boldsymbol{y}} = J_N \nabla H(\boldsymbol{y}), \qquad \boldsymbol{y}(0) = \begin{pmatrix} \psi_0(\boldsymbol{x}) \\ \psi_1(\boldsymbol{x}) \end{pmatrix} =: \boldsymbol{y}_0, \tag{5.23}$$

with Hamiltonian:

$$H = H(\boldsymbol{q}, \boldsymbol{p}) := \Delta x \left[\frac{\boldsymbol{p}^\top \boldsymbol{p}}{2} + \frac{\boldsymbol{q}^\top T_N \boldsymbol{q}}{2 \Delta x^2} + \mathbf{1}^\top f(\boldsymbol{q}) \right]. \tag{5.24}$$

Writing (5.24) in componentwise form,

$$H(\boldsymbol{q}, \boldsymbol{p}) = \Delta x \sum_{i=0}^{N-1} \left(\frac{1}{2} v_i^2 - u_i \frac{u_{i-1} - 2u_i + u_{i+1}}{2\Delta x^2} + f(u_i) \right), \qquad (5.25)$$

we see that (5.24) is nothing but the approximation of (5.16) via the composite trapezoidal rule (provided that the second derivative u_{xx} is approximated as indicated at (5.18), and taking into account the periodic boundary conditions (5.2)). Consequently, we can conclude that (5.25) is a $O(\Delta x^2)$ approximation to (5.16). The Hamiltonian (5.25) is time independent, so it is a conserved quantity for (5.23), similarly to what happens in the continuous case for (5.16). It follows that HBVMs can be conveniently used for solving (5.23). In particular, the following result is easily established, by using arguments similar to those used in Section 3.1.1 (see Theorems 3.4 and 3.5 on pages 91 and 92, respectively).

Theorem 5.2. *Assume $k \geqslant s$, and define $\boldsymbol{y}_1$ as the approximation to $\boldsymbol{y}(h)$ yielded by a HBVM(k, s) method used with stepsize h for solving (5.23). One then obtains:*

$$\boldsymbol{y}_1 - \boldsymbol{y}(h) = O(h^{2s+1}),$$

that is the method has order $2s$. Moreover, with reference to (5.24), and assuming that f is suitably regular:

$$H(\boldsymbol{y}_1) - H(\boldsymbol{y}_0) = \begin{cases} 0, & \text{if } f \in \Pi_\nu \text{ and } \nu \leqslant 2k/s, \\ O(h^{2k+1}), & \text{otherwise.} \end{cases}$$

5.2.1   Higher-order space discretization

In the case where periodic conditions are prescribed for the problem, one can use any symmetric discretization for the second space derivative in place of (5.18). As an example, the following fourth-order approximation,

$$u_{xx}(x_i, t) \approx \frac{1}{\Delta x^2} \Big(-\frac{1}{12} u(x_{i+2}, t)$$
$$+ \frac{4}{3} u(x_{i+1}, t) - \frac{5}{2} u(x_i, t) + \frac{4}{3} u(x_{i-1}, t) - \frac{1}{12} u(x_{i-2}, t) \Big),$$

allows us to replace (5.1) with the following set of differential equations,

$$\ddot{u}_i(t) = \frac{1}{\Delta x^2} \Big(-\frac{1}{12} u_{i+2}(t) + \frac{4}{3} u_{i+1}(t) - \frac{5}{2} u_i(t) \qquad (5.26)$$
$$+ \frac{4}{3} u_{i-1}(t) - \frac{1}{12} u_{i-2}(t) \Big) - f'(u_i), \qquad i = 0, \dots, N-1, \qquad t \geqslant 0,$$

where, because of the periodic boundary conditions,

$$u_{N+1}(t) = u_1(t), \quad u_N(t) = u_0(t), \quad u_{-1}(t) = u_{N-1}(t), \quad u_{-2}(t) = u_{N-2}(t).$$

As a consequence, the Hamiltonian of the resulting semi-discrete problem, given by (5.26) and (5.20), turns out to be formally still equal to (5.24), with matrix T_N in (5.21) replaced by

$$T_N = \begin{pmatrix} \frac{5}{2} & -\frac{4}{3} & \frac{1}{12} & & & \frac{1}{12} & -\frac{4}{3} \\ -\frac{4}{3} & \ddots & \ddots & \ddots & & & \frac{1}{12} \\ \frac{1}{12} & \ddots & \ddots & \ddots & \ddots & & \\ & \ddots & \ddots & \ddots & \ddots & \ddots & \\ & & \ddots & \ddots & \ddots & \ddots & \frac{1}{12} \\ \frac{1}{12} & & & \ddots & \ddots & \ddots & -\frac{4}{3} \\ -\frac{4}{3} & \frac{1}{12} & & & \frac{1}{12} & -\frac{4}{3} & \frac{5}{2} \end{pmatrix} \in \mathbb{R}^{N \times N}.$$

In such a way, one obtains a fourth-order approximation to the Hamiltonian (5.16).[4] Further higher-order approximations can be obtained by considering different symmetric and circulant matrices T_N.

5.2.2 Fourier space expansion

An alternative approach to the problem of obtaining higher-order space discretizations is that of using a Fourier approximation. For this purpose, let us consider the following complete set of orthonormal functions on $[0, 1]$:

$$c_0(x) \equiv 1, \quad c_k(x) = \sqrt{2}\cos(2k\pi x), \quad s_k(x) = \sqrt{2}\sin(2k\pi x), \quad k = 1, 2, \ldots, \tag{5.27}$$

so that

$$\int_0^1 c_i(x)c_j(x)\mathrm{d}x = \int_0^1 s_i(x)s_j(x)\mathrm{d}x = \delta_{ij}, \quad \int_0^1 c_i(x)s_j(x)\mathrm{d}x = 0, \quad \forall i, j. \tag{5.28}$$

The following expansion of the solution of (5.1)-(5.2) is a slightly different way of writing the usual Fourier expansion in space:

$$u(x, t) = c_0(x)\gamma_0(t) + \sum_{n \geqslant 1} [c_n(x)\gamma_n(t) + s_n(x)\eta_n(t)]$$

$$\equiv \gamma_0(t) + \sum_{n \geqslant 1} [c_n(x)\gamma_n(t) + s_n(x)\eta_n(t)], \quad x \in [0, 1], \quad t \geqslant 0, \tag{5.29}$$

with

$$\gamma_n(t) = \int_0^1 c_n(x)u(x, t)\mathrm{d}x, \qquad \eta_n(t) = \int_0^1 s_n(x)u(x, t)\mathrm{d}x,$$

---

[4]In fact, the composite trapezoidal rule converges more than exponentially in N, if the argument is a suitably regular periodic function.

which is allowed because of the periodic boundary conditions (5.2). Consequently, by taking into account (5.28), the first equation in (5.1) can be rewritten as

$$\ddot{\gamma}_n(t) = -(2\pi n)^2 \gamma_n(t)$$
$$- \int_0^1 c_n(x) f'\left(\gamma_0(t) + \sum_{j\geq 1}[c_j(x)\gamma_j(t) + s_j(x)\eta_j(t)]\right) dx, \quad n \geq 0,$$

(5.30)

$$\ddot{\eta}_n(t) = -(2\pi n)^2 \eta_n(t)$$
$$- \int_0^1 s_n(x) f'\left(\gamma_0(t) + \sum_{j\geq 1}[c_j(x)\gamma_j(t) + s_j(x)\eta_j(t)]\right) dx, \quad n \geq 1.$$

The initial conditions for (5.30) are clearly given by (see (5.1)):

$$\gamma_n(0) = \int_0^1 c_n(x)\psi_0(x)dx, \qquad \eta_n(0) = \int_0^1 s_n(x)\psi_0(x)dx,$$

(5.31)

$$\dot{\gamma}_n(0) = \int_0^1 c_n(x)\psi_1(x)dx, \qquad \dot{\eta}_n(0) = \int_0^1 s_n(x)\psi_1(x)dx.$$

Introducing the infinite vectors

$$\boldsymbol{\omega}(x) = \begin{pmatrix} c_0(x), & c_1(x), & s_1(x), & c_2(x), & s_2(x), & \cdots \end{pmatrix}^\top,$$

(5.32)

$$\boldsymbol{q}(t) = \begin{pmatrix} \gamma_0(t), & \gamma_1(t), & \eta_1(t), & \gamma_2(t), & \eta_2(t), & \cdots \end{pmatrix}^\top,$$

the infinite matrix

$$D = \begin{pmatrix} 0 & & & & & \\ & (2\pi)^2 & & & & \\ & & (2\pi)^2 & & & \\ & & & (4\pi)^2 & & \\ & & & & (4\pi)^2 & \\ & & & & & \ddots \end{pmatrix},$$

(5.33)

and considering that (see (5.29))

$$u(x,t) = \boldsymbol{\omega}(x)^\top \boldsymbol{q}(t),$$

(5.34)

problem (5.30) can be cast in vector form as:

$$\dot{\boldsymbol{q}}(t) = \boldsymbol{p}(t), \qquad t \geq 0,$$

(5.35)

$$\dot{\boldsymbol{p}}(t) = -D\boldsymbol{q}(t) - \int_0^1 \boldsymbol{\omega}(x) f'(\boldsymbol{\omega}(x)^\top \boldsymbol{q}(t))\, dx,$$

with the initial conditions (5.31) written, more compactly, as

$$q(0) = \int_0^1 \boldsymbol{\omega}(x)\psi_0(x)\,\mathrm{d}x, \qquad p(0) = \int_0^1 \boldsymbol{\omega}(x)\psi_1(x)\,\mathrm{d}x.$$

The following result then holds true.

Theorem 5.3. *Problem (5.35) is Hamiltonian, with Hamiltonian*

$$H(\boldsymbol{q},\boldsymbol{p}) = \frac{1}{2}\boldsymbol{p}^\top\boldsymbol{p} + \frac{1}{2}\boldsymbol{q}^\top D\boldsymbol{q} + \int_0^1 f(\boldsymbol{\omega}(x)^\top\boldsymbol{q})\,\mathrm{d}x. \tag{5.36}$$

This latter is equivalent to the Hamiltonian functional (5.7), via the expansion (5.29)–(5.34).

Proof. The first statement is straightforward, by considering that

$$\nabla_{\boldsymbol{q}} f(\boldsymbol{\omega}(x)^\top\boldsymbol{q})) = f'(\boldsymbol{\omega}(x)^\top\boldsymbol{q})\boldsymbol{\omega}(x).$$

Taking into account (5.34), the second statement easily follows from the fact that (see (5.28), (5.29), and (5.32))

$$
\begin{aligned}
\int_0^1 v(x,t)^2\mathrm{d}x &= \int_0^1 u_t(x,t)^2\mathrm{d}x \\
&= \int_0^1 \left(\dot{\gamma}_0(t) + \sum_{j\geqslant 1}[\dot{\gamma}_j(t)c_j(x) + \dot{\eta}_j(t)s_j(x)] \right)^2 \mathrm{d}x \\
&= \dot{\gamma}_0(t)^2 + \sum_{j\geqslant 1}[\dot{\gamma}_j(t)^2 + \dot{\eta}_j(t)^2] \;=\; \boldsymbol{p}(t)^\top\boldsymbol{p}(t),
\end{aligned}
$$

and

$$
\begin{aligned}
\int_0^1 u_x(x,t)^2\mathrm{d}x &= \int_0^1 \left(\sum_{j\geqslant 1} 2\pi j\,[\eta_j(t)c_j(x) - \gamma_j(t)s_j(x)] \right)^2 \mathrm{d}x \\
&= \sum_{j\geqslant 1}(2\pi j)^2\,[\eta_j(t)^2 + \gamma_j(t)^2] \;=\; \boldsymbol{q}(t)^\top D\boldsymbol{q}(t).
\end{aligned}
$$

$\square$

5.2.3 Fourier-Galerkin space discretization

In order to obtain a practical computational procedure, we truncate the infinite expansion (5.29) to a finite sum:

$$u(x,t) \approx \gamma_0(t) + \sum_{n=1}^N [c_n(x)\gamma_n(t) + s_n(x)\eta_n(t)] \;=:\; u_N(x,t), \tag{5.37}$$

which is known to converge more than exponentially with N to u, if this latter is an analytical function. In other words, we look for an approximation to $u(x, t)$ belonging to the functional subspace (see (5.27))

$$\mathcal{V}_N = \text{span}\left\{c_0(x), c_1(x), \ldots, c_N(x), s_1(x), \ldots, s_N(x)\right\}.$$

Clearly, such a truncated expansion will not satisfy problem (5.1)-(5.2). Nevertheless, in the spirit of Fourier-Galerkin methods, by requiring that the residual

$$R(u_N) := (u_N)_{tt} - (u_N)_{xx} + f'(u_N)$$

be orthogonal to $\mathcal{V}_N$, one obtains the *weak formulation* of problem (5.1)-(5.2), consisting of the following set of $2N + 1$ differential equations,

$$\ddot{\gamma}_n(t) = -(2\pi n)^2 \gamma_n(t)$$
$$- \int_0^1 c_n(x) f'\left(\gamma_0(t) + \sum_{j=1}^{N} [c_j(x)\gamma_j(t) + s_j(x)\eta_j(t)]\right) dx, \quad n = 0, \ldots, N,$$

$$(5.38)$$

$$\ddot{\eta}_n(t) = -(2\pi n)^2 \eta_n(t)$$
$$- \int_0^1 s_n(x) f'\left(\gamma_0(t) + \sum_{j=1}^{N} [c_j(x)\gamma_j(t) + s_j(x)\eta_j(t)]\right) dx, \quad n = 1, \ldots, N,$$

approximating the leading ones in (5.30). Defining the finite vectors in $\mathbb{R}^{2N+1}$ (compare with (5.32)),

$$\boldsymbol{\omega}_N(x) = \begin{pmatrix} c_0(x), & c_1(x), & s_1(x), & \ldots, & c_N(x), & s_N(x) \end{pmatrix}^\top,$$

$$\boldsymbol{q}_N(t) = \begin{pmatrix} \gamma_0(t), & \gamma_1(t), & \eta_1(t), & \ldots, & \gamma_N(t), & \eta_N(t) \end{pmatrix}^\top,$$

the matrix in $\mathbb{R}^{(2N+1)\times(2N+1)}$ (compare with (5.33))

$$D_N = \begin{pmatrix} 0 & & & & & & \\ & (2\pi)^2 & & & & & \\ & & (2\pi)^2 & & & & \\ & & & (4\pi)^2 & & & \\ & & & & (4\pi)^2 & & \\ & & & & & \ddots & \\ & & & & & & (2N\pi)^2 & \\ & & & & & & & (2N\pi)^2 \end{pmatrix},$$

and considering that (compare with (5.34))

$$u_N(x, t) = \boldsymbol{\omega}_N(x)^\top \boldsymbol{q}_N(t), \quad (5.39)$$

equations (5.38) can be cast in vector form as

$$\dot{\boldsymbol{q}}_N(t) = \boldsymbol{p}_N(t), \qquad t \geqslant 0, \tag{5.40}$$

$$\dot{\boldsymbol{p}}_N(t) = -D_N \boldsymbol{q}_N(t) - \int_0^1 \boldsymbol{\omega}_N(x) f'(\boldsymbol{\omega}_N(x)^\top \boldsymbol{q}_N(t)) \mathrm{d}x,$$

for a total of $4N + 2$ differential equations. From (5.31) we see that the initial conditions for (5.40) are given by

$$\boldsymbol{q}_N(0) = \int_0^1 \boldsymbol{\omega}_N(x) \psi_0(x) \mathrm{d}x, \qquad \boldsymbol{p}_N(0) = \int_0^1 \boldsymbol{\omega}_N(x) \psi_1(x) \mathrm{d}x.$$

The following result then easily follows by means of arguments similar to those used to prove Theorem 5.3.

Theorem 5.4. *Problem (5.40) is Hamiltonian, with Hamiltonian*

$$H_N(\boldsymbol{q}_N, \boldsymbol{p}_N) = \frac{1}{2} \boldsymbol{p}_N^\top \boldsymbol{p}_N + \frac{1}{2} \boldsymbol{q}_N^\top D_N \boldsymbol{q}_N + \int_0^1 f(\boldsymbol{\omega}_N(x)^\top \boldsymbol{q}_N) \mathrm{d}x. \tag{5.41}$$

We observe that (5.41) is the semi-discrete analogue of the Hamiltonian (5.36). However, it must be stressed that, unlike the finite-difference case, $H_N \to H$ more than exponentially in N, provided that the involved functions are analytical.

Again, since (5.40) represents a canonical Hamiltonian system, energy conserving methods can be conveniently used to determine numerical solutions along which the Hamiltonian function (5.41) is precisely conserved. In particular, Theorem 5.2 continues formally to hold in this context: the HBVM(k, s) has order $2s$ in time, and the error in the semi-discrete Hamiltonian (5.41) is of order $2k$, for all $k \geqslant s$.[5]

The integral appearing in (5.40) must be approximated by means of a suitable quadrature rule. To this end, it could be convenient to employ a composite trapezoidal rule, due to the fact that the integrand function is periodic. Setting

$$\boldsymbol{g}_N(x, t) = \boldsymbol{\omega}_N(x) f'(\boldsymbol{\omega}_N(x)^\top \boldsymbol{q}_N(t)), \tag{5.42}$$

introducing the uniform mesh on $[0, 1]$

$$x_i = i\Delta x, \qquad i = 0, \ldots, m, \qquad \Delta x = \frac{1}{m}, \tag{5.43}$$

and considering that $\boldsymbol{g}_N(0, t) = \boldsymbol{g}_N(1, t)$, we obtain

$$\int_0^1 \boldsymbol{g}_N(x, t) \mathrm{d}x = \Delta x \sum_{i=1}^m \frac{\boldsymbol{g}_N(x_{i-1}, t) + \boldsymbol{g}_N(x_i, t)}{2} + R(m)$$

$$= \frac{1}{m} \sum_{i=0}^{m-1} \boldsymbol{g}_N(x_i, t) + R(m). \tag{5.44}$$

---

[5]Notice that here the integrands are no longer polynomials.

To study the quadrature error $R(m)$, we will make use of the following well-known result.

Lemma 5.1. *Let us consider the trigonometric polynomial*

$$p(x) = a_0 + \sum_{k=1}^{K} [a_k \cos(2k\pi x) + b_k \sin(2k\pi x)], \qquad (5.45)$$

and the uniform mesh (5.43). Then, for all $m \geqslant K+1$, the following property holds true:

$$\int_0^1 p(x)\mathrm{d}x = \frac{1}{m} \sum_{i=0}^{m-1} p(x_i).$$

Lemma 5.2. *Let us consider the trigonometric polynomial (5.45) and the uniform mesh (5.43). Then, for all $m \geqslant N + K + 1$,*

$$\int_0^1 \cos(2j\pi x)p(x)\mathrm{d}x = \frac{1}{m} \sum_{i=0}^{m-1} \cos(2j\pi x_i)p(x_i), \qquad (5.46)$$

$$\int_0^1 \sin(2j\pi x)p(x)\mathrm{d}x = \frac{1}{m} \sum_{i=0}^{m-1} \sin(2j\pi x_i)p(x_i), \quad j = 0, \dots, N.$$

Proof. By virtue of the prosthaphaeresis formulae, one has, for all $j = 0, \dots, N$ and $k = 0, \dots, K$:

$$\cos(2j\pi x)\cos(2k\pi x) = \frac{1}{2}\left[\cos(2(k+j)\pi x) + \cos(2(k-j)\pi x)\right],$$

$$\cos(2j\pi x)\sin(2k\pi x) = \frac{1}{2}\left[\sin(2(k+j)\pi x) + \sin(2(k-j)\pi x)\right],$$

$$\sin(2j\pi x)\cos(2k\pi x) = \frac{1}{2}\left[\sin(2(k+j)\pi x) - \sin(2(k-j)\pi x)\right],$$

$$\sin(2j\pi x)\sin(2k\pi x) = \frac{1}{2}\left[\cos(2(k-j)\pi x) - \cos(2(k+j)\pi x)\right].$$

Consequently, the integrals at the left-hand sides in (5.46) are trigonometric polynomials of degree at most $N + K$. By virtue of Lemma 5.1, it then follows that they are exactly computed by means of the composite trapezoidal rule at the corresponding right-hand sides, provided that $m \geqslant N + K + 1$. ∎

By virtue of Lemma 5.2, the following result is readily established.

Theorem 5.5. *Let the function f appearing in (5.42) be a polynomial of degree ν, and let us consider the uniform mesh (5.43). Then, with reference to (5.44), for all $m \geqslant \nu N + 1$ one obtains:*

$$R(m) = 0 \qquad i.e., \qquad \int_0^1 \mathbf{g}_N(x,t)\mathrm{d}x = \frac{1}{m} \sum_{i=0}^{m-1} \mathbf{g}_N(x_i,t).$$

For a general function f, it can be proved that $R(m) = O(m^{-r})$, if f has r continuous derivatives (as a periodic function). Consequently, the convergence is more than exponential, for C^∞ periodic functions (in such a case, one speaks of a *spectrally accurate* approximation.)

5.3 Nonperiodic boundary conditions

Let us now consider the case where the prescribed boundary conditions are not periodic, i.e., we are solving problem (5.1) with either Dirichlet (5.4) or Neumann (5.5) boundary conditions. Consequently, the continuous Hamiltonian is given by (5.15), and its derivative is given by (5.14). To solve numerically such a problem we shall resort again to a finite-difference space semi-discretization, thus considering the following set of discrete points,

$$x_i = i\Delta x, \qquad i = 0, \dots, N+1, \qquad \Delta x = 1/(N+1),$$

with corresponding approximations $u_i(t) \approx u(x_i, t)$ and $v_i(t) := \dot{u}_i(t) \approx v(x_i, t)$. Clearly, such approximations have to be computed only at the inner grid-points x_i, $i = 1, \dots, N$, since $u_0(t)$ and $u_{N+1}(t)$ will be obtained through the prescribed boundary conditions.

To derive a semi-discrete version of the Hamiltonian (5.15), we formally replace the first space derivatives at the boundary points as

$$u_x(1, t) \to \frac{u_{N+1}(t) - u_N(t)}{\Delta x}, \qquad u_x(0, t) \to \frac{u_1(t) - u_0(t)}{\Delta x},$$

meanwhile replacing the second space derivatives at the inner grid-points as

$$u_{xx}(x_i, t) \to \frac{u_{i+1}(t) - 2u_i(t) + u_{i-1}(t)}{\Delta x^2}, \qquad i = 1, \dots, N.$$

In so doing, we obtain the following approximation to (5.15),

$$H = \Delta x \sum_{i=1}^N \left(\frac{1}{2} v_i^2 - u_i \frac{u_{i-1} - 2u_i + u_{i+1}}{2\Delta x^2} + f(u_i) \right)$$
$$+ \frac{1}{2} \left[u_{N+1} \frac{u_{N+1} - u_N}{\Delta x} - u_0 \frac{u_1 - u_0}{\Delta x} \right],$$

which can be also written in the equivalent form:

$$H = \Delta x \left[\sum_{i=2}^{N-1} \left(\frac{1}{2} v_i^2 - u_i \frac{u_{i-1} - 2u_i + u_{i+1}}{2\Delta x^2} + f(u_i) \right) \right.$$
$$+ \frac{1}{2} v_1^2 - u_1 \frac{-u_1 + u_2}{2\Delta x^2} + f(u_1) + \frac{1}{2} v_N^2 - u_N \frac{u_{N-1} - u_N}{2\Delta x^2} + f(u_N) \right]$$
$$+ \frac{(u_1 - u_0)^2}{2\Delta x} + \frac{(u_{N+1} - u_N)^2}{2\Delta x}. \tag{5.47}$$

Defining the vectors in $\mathbb{R}^N$,[6]

$$q = \begin{pmatrix} u_1 \\ \vdots \\ u_N \end{pmatrix}, \quad p = \begin{pmatrix} v_1 \\ \vdots \\ v_N \end{pmatrix}, \quad w(q,t) = \begin{pmatrix} u_1 - u_0(t) \\ 0 \\ \vdots \\ 0 \\ u_{N+1}(t) - u_N \end{pmatrix}, \quad (5.48)$$

and the matrix

$$\bar{T}_N = \begin{pmatrix} 1 & -1 & & & \\ -1 & 2 & \ddots & & \\ & \ddots & \ddots & \ddots & \\ & & \ddots & 2 & -1 \\ & & & -1 & 1 \end{pmatrix} \in \mathbb{R}^{N \times N},$$

we can rewrite (5.47) as

$$H = H(q,p,t) := \Delta x \left[\frac{p^\top p}{2} + \frac{q^\top \bar{T}_N q}{2\Delta x^2} + 1^\top f(q) \right] + \frac{w(q,t)^\top w(q,t)}{2\Delta x}. \quad (5.49)$$

Correspondingly, one derives the following Hamiltonian problem

$$\dot{q} = p \equiv \frac{1}{\Delta x} \nabla_p H(q,p,t), \qquad t \geqslant 0, \qquad\qquad (5.50)$$

$$\dot{p} = -\frac{1}{\Delta x^2} \left[\bar{T}_N q + \frac{\partial}{\partial q} w(q,t) \, w(q,t) \right] - f'(q) \equiv \frac{-1}{\Delta x} \nabla_q H(q,p,t),$$

with initial conditions given by

$$q(0) = \psi_0(x), \qquad p(0) = \psi_1(x), \qquad x = \begin{pmatrix} x_1, & \dots, & x_N \end{pmatrix}^\top. \quad (5.51)$$

In fact, in componentwise form (5.50) reads

$$\dot{u}_i = v_i, \qquad \dot{v}_i = \frac{u_{i-1} - 2u_i + u_{i+1}}{\Delta x^2} - f'(u_i), \qquad i = 1, \dots, N,$$

with u_0 and u_{N+1} obtained through the prescribed boundary conditions, i.e.,

$$u_0(t) = \varphi_0(t), \qquad u_{N+1}(t) = \varphi_1(t), \qquad t \geqslant 0, \qquad (5.52)$$

when the Dirichlet boundary conditions (5.4) are selected, or

$$u_0(t) = u_1(t) - \Delta x \, \phi_0(t), \qquad u_{N+1}(t) = u_N(t) + \Delta x \, \phi_1(t), \qquad t \geqslant 0, \quad (5.53)$$

---

[6]We have specified the formal argument t for w since u_0 and u_{N+1} will depend on the prescribed boundary conditions, which are functions of t.

when the Neumann boundary conditions (5.5) are considered.

We observe that the discrete Hamiltonian is in general no longer conserved in this case. In fact,

$$\frac{\mathrm{d}}{\mathrm{d}t}H(\boldsymbol{q},\boldsymbol{p},t) = \overbrace{\nabla_{\boldsymbol{q}}H(\boldsymbol{q},\boldsymbol{p},t)^{\top}\dot{\boldsymbol{q}} + \nabla_{\boldsymbol{p}}H(\boldsymbol{q},\boldsymbol{p},t)^{\top}\dot{\boldsymbol{p}}}^{=\,0,\ \text{because of (5.50)}} + \frac{\partial}{\partial t}H(\boldsymbol{q},\boldsymbol{p},t)$$

$$= \frac{\partial}{\partial t}H(\boldsymbol{q},\boldsymbol{p},t) = \frac{1}{\Delta x}\boldsymbol{w}(\boldsymbol{q},t)^{\top}\frac{\partial}{\partial t}\boldsymbol{w}(\boldsymbol{q},t) \qquad (5.54)$$

$$= \frac{u_{N+1}-u_{N}}{\Delta x}\dot{u}_{N+1} - \frac{u_{1}-u_{0}}{\Delta x}\dot{u}_{0}$$

$$= \begin{cases} \dfrac{\varphi_{1}(t)-u_{N}}{\Delta x}\dot{\varphi}_{1}(t) - \dfrac{u_{1}-\varphi_{0}(t)}{\Delta x}\dot{\varphi}_{0}(t), & \text{if (5.52) holds,} \\[2mm] \phi_{1}(t)\left(v_{N}+\Delta x\,\dot{\phi}_{1}(t)\right) - \phi_{0}(t)\left(v_{1}-\Delta x\,\dot{\phi}_{0}(t)\right), & \text{if (5.53) holds.} \end{cases}$$

Consequently, the discrete Hamiltonian is conserved only when constant Dirichlet boundary conditions (so that $\dot{\varphi}_{1}(t) = \dot{\varphi}_{0}(t) \equiv 0$) or homogeneous Neumann boundary conditions (i.e., $\phi_{1}(t) = \phi_{0}(t) \equiv 0$) are prescribed: for these specific events, H turns out to be autonomous, in agreement with cases **ii)** and **iii)** listed on page 149.

In the general non-autonomous case, we can "embed" the Hamiltonian system defined by (5.49) in an extended phase space, by introducing a couple of auxiliary scalar conjugate variables $\tilde{q}$ and $\tilde{p}$, that make the resulting augmented system autonomous. More precisely, associated with (5.49), we consider the new Hamiltonian function

$$\widetilde{H}(\boldsymbol{q},\boldsymbol{p},\tilde{q},\tilde{p}) := H(\boldsymbol{q},\boldsymbol{p},\tilde{q}) + \tilde{p}. \qquad (5.55)$$

The dynamical system corresponding to (5.55) is then given by

$$\begin{aligned} \dot{\boldsymbol{q}} &= \frac{1}{\Delta x}\nabla_{\boldsymbol{p}}\widetilde{H} \equiv \frac{1}{\Delta x}\nabla_{\boldsymbol{p}}H, \\[1mm] \dot{\boldsymbol{p}} &= -\frac{1}{\Delta x}\nabla_{\boldsymbol{q}}\widetilde{H} \equiv -\frac{1}{\Delta x}\nabla_{\boldsymbol{q}}H, \qquad t \geqslant 0, \\[1mm] \frac{\mathrm{d}}{\mathrm{d}t}\tilde{q} &= 1 \equiv \frac{\partial}{\partial \tilde{p}}\widetilde{H}, \qquad\qquad\qquad (5.56) \\[1mm] \frac{\mathrm{d}}{\mathrm{d}t}\tilde{p} &= -\frac{\partial}{\partial \tilde{q}}\widetilde{H} \equiv -\frac{\partial}{\partial t}H\Big|_{t=\tilde{q}}, \end{aligned}$$

with H and $\frac{\partial}{\partial t}H$ defined at (5.49) and (5.54), respectively. Using the initial conditions (5.51) and

$$\tilde{q}(0) = \tilde{p}(0) = 0,$$

we see that the first two equations in (5.56) coincide with (5.50), whereas the variables $\tilde{q}(t) \equiv t$ and (see the last equation in (5.56))

$$\tilde{p} \equiv H\big|_{t=0} - H\big|_{t=\tilde{q}} \qquad (5.57)$$

allow for energy conservation. In fact, by setting

$$\boldsymbol{y} := \begin{pmatrix} \boldsymbol{q} \\ \boldsymbol{p} \\ \tilde{q} \\ \tilde{p} \end{pmatrix} \in \mathbb{R}^{2N+2}, \qquad \tilde{J}_N = \frac{1}{\Delta x} \left(\begin{array}{cc|cc} & I_N & & \\ -I_N & & & \\ \hline & & & \Delta x \\ & & -\Delta x & \end{array} \right),$$

one obtains that (5.56) can be written, more compactly, as

$$\dot{\boldsymbol{y}} = \tilde{J}_N \nabla \tilde{H}(\boldsymbol{y}), \quad t \geqslant 0, \qquad \boldsymbol{y}(0) = \boldsymbol{y}_0, \tag{5.58}$$

and, since $\tilde{J}_N$ is skew-symmetric, one has:

$$\frac{\mathrm{d}}{\mathrm{d}t} \tilde{H}(\boldsymbol{y}) = \nabla \tilde{H}(\boldsymbol{y})^\top \tilde{J}_N \nabla \tilde{H}(\boldsymbol{y}) = 0.$$

Consequently,

$$\tilde{H}(\boldsymbol{y}(t)) = \tilde{H}(\boldsymbol{y}_0) \equiv H(\boldsymbol{q}(0), \boldsymbol{p}(0), 0), \qquad \forall t \geqslant 0.$$

Since $\tilde{H}$ is conserved for problem (5.58), HBVMs can be conveniently used for its solution. In particular, the following result holds true.

Theorem 5.6. *Assume $k \geqslant s$, and define $\boldsymbol{y}_1$ as the new approximation to $\boldsymbol{y}(h)$, solution of (5.58), provided by a HBVM(k, s) method used with stepsize h. Then*

$$\boldsymbol{y}_1 - \boldsymbol{y}(h) = O(h^{2s+1}),$$

that is the method has order $2s$. Moreover, assuming that f is suitably regular, as well as φ_0 and φ_1, when the Dirichlet boundary conditions (5.4) are prescribed, or ϕ_0 and ϕ_1, when the Neumann boundary conditions (5.5) are prescribed, one has:

$$\tilde{H}(\boldsymbol{y}_1) - \tilde{H}(\boldsymbol{y}_0) = \begin{cases} 0, & \textit{if } f \in \Pi_\nu, \textit{ the boundary conditions} \\ & \textit{(5.4) are prescribed, } \varphi_0, \varphi_1 \in \Pi_\rho, \textit{ with} \\ & k \geqslant \frac{1}{2} \max\{\nu s, 2\rho + s - 1, \rho + 2s - 1\}, \\[2ex] 0, & \textit{if } f \in \Pi_\nu, \textit{ the boundary conditions} \\ & \textit{(5.5) are prescribed, } \phi_0, \phi_1 \in \Pi_\rho, \textit{ with} \\ & 2k \geqslant \max\{\nu s, 2\rho + s - 1, 2s + \rho\}, \\[2ex] O(h^{2k+1}), & \textit{otherwise.} \end{cases}$$

Remark

From (5.57) we see that the variable $\tilde{p}$ measures the variation of the original Hamiltonian function H along the solution as time progresses. Looking at (5.55) we argue that conserving the augmented Hamiltonian $\tilde{H}$ while integrating the system numerically, may improve the ability of the method in reproducing the correct variation of H in time.

Moreover, in the case where $|\tilde{p}| \simeq \varepsilon \ll |\tilde{H}|$ (see (5.55)), one may expect the error on the semi-discrete Hamiltonian $H \equiv \tilde{H} - \tilde{p}$ to behave as

$$O(h^{2k}) + \varepsilon O(h^{2s}). \tag{5.59}$$

Consequently, the error on H will approximately decrease with order $2k$, until $O(h^{2k}) \simeq \varepsilon O(h^{2s})$. This aspect will be confirmed by the numerical tests in the next section.

5.4 Numerical tests

We here report a few numerical tests concerning the *Sine-Gordon* equation, which is in the form (5.1):

$$u_{tt}(x,t) = u_{xx}(x,t) - \sin(u(x,t)), \qquad x \in (-20, 20), \quad t \geqslant 0. \tag{5.60}$$

In particular, we shall consider special *soliton-like* solutions, defined by the initial conditions:

$$u(x,0) \equiv 0, \qquad u_t(x,0) = \frac{4}{\gamma}\operatorname{sech}\left(\frac{x}{\gamma}\right), \qquad \gamma > 0. \tag{5.61}$$

Depending on the value of the positive parameter γ, the solution is known to be given by:

$$u(x,t) = 4\operatorname{atan}\left[\varphi(t;\gamma)\operatorname{sech}\left(\frac{x}{\gamma}\right)\right],$$

with

$$\varphi(t;\gamma) = \begin{cases} (\sqrt{\gamma^2-1})^{-1}\sin\left(\gamma^{-1}\sqrt{\gamma^2-1}\,t\right), & \text{if} \quad \gamma > 1, \\ t, & \text{if} \quad \gamma = 1, \\ (\sqrt{1-\gamma^2})^{-1}\sinh\left(\gamma^{-1}\sqrt{1-\gamma^2}\,t\right), & \text{if} \quad 0 < \gamma < 1. \end{cases}$$

The three cases are shown in Figure 5.1: in the upper-left plot one sees the first soliton (obtained for $\gamma > 1$), which is named *breather*; in the upper-right plot is the case $0 < \gamma < 1$, which is named *kink-antikink*; at last, the case $\gamma = 1$, which is named *double-pole*, separates the two different types of dynamics and is shown in the lower-left plot. Moreover, the space interval being fixed,[7] the Hamiltonian is a decreasing function of γ, as is shown in the lower-right plot of Figure 5.1. This means that the value of the Hamiltonian characterizes the dynamics. Consequently, in a neighborhood of $\gamma = 1$, where the Hamiltonian assumes a value $\simeq 16$, nearby values of the Hamiltonian will provide different types of soliton solutions. As a result, energy conserving methods are expected to be useful, when numerically solving problem (5.60)-(5.61) with $\gamma = 1$.

To begin with, we consider the case where system (5.60)-(5.61), with $\gamma = 1$, is coupled with periodic boundary conditions. We perform both a finite difference space discretization, with $N + 1 = 401$ equally spaced mesh points (i.e., $N = 400$ in (5.17)), and a Fourier-Galerkin discretization, with $N = 100$ Fourier modes and $m + 1 = 201$ grid-points for approximating the integrals (i.e., $m = 200$ in (5.43)).

In both cases, we use the symplectic implicit mid-point method (i.e., HBVM(1,1)) and the (practically) energy-conserving method HBVM(5,1), with stepsize $h = 10^{-1}$, for 10^3 time-steps. The obtained results are summarized in Figure 5.2: on the left are the results for the finite-difference space discretization, whereas those for the Fourier-Galerkin space discretization are on the right. In more detail:

- the upper pictures display the errors in the numerical Hamiltonian function, which is not negligible for HBVM(1,1), whereas it is of the order of round-off for HBVM(5,1);

- in the middle pictures are the solutions computed by HBVM(1,1). Evidently, they are not correct, since the double-pole has become a breather-type soliton;

- in the lower pictures are the solutions computed by HBVM(5,1). Their shape is correctly reproduced, as one infers by comparing them with the left-lower pictures in Figure 5.1.

Similar results are obtained when the problem is coupled with Dirichlet or Neumann boundary conditions, for which the Hamiltonian (5.49) is no longer conserved. As an example, by prescribing Dirichlet boundary conditions, and using a finite-difference space discretization as above, the two methods HBVM(1,1) and HBVM(5,1), applied with stepsize $h = 10^{-1}$, yield the results presented in Figure 5.3, and explained below.

- The plots on the left concern HBVM(1,1): in the upper plot, is the difference of the Hamiltonian H in (5.49) from its initial value, whereas in the intermediate plot is the difference of the augmented Hamiltonian $\tilde{H}$ (5.55) from its initial value. As is clear, both of them are not conserved, and yield quite similar values. In the bottom picture is the computed solution, which is wrong, since it is of breather-type.

---

[7]I.e., $(-20, 20)$, in our case (see (5.60)).

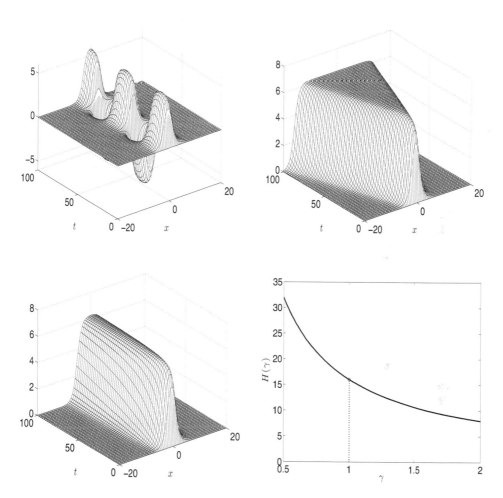

FIGURE 5.1: Sine-Gordon problem (5.60)-(5.61). Upper-left: *breather* soliton ($\gamma > 1$). Upper-right: *kink-antikink* soliton ($0 < \gamma < 1$). Lower-left: *double-pole* soliton ($\gamma = 1$). Lower-right: Hamiltonian as a function of γ.

- The pictures on the right concern HBVM(5, 1) and show the same functions we have described above. Now the augmented Hamiltonian function is conserved to within machine precision while the variation in time of the Hamiltonian H in (5.49) is quite well represented. In the lower plot is the computed solution, which is now qualitatively correct (compare with the left-lower plot in Figure 5.1).

As a further illustration of the behavior of the HBVM(k, 1) family of methods for the problem at hand, we wish to investigate how the choice of the parameter k may influence the correct reproduction of the time varying Hamiltonian function H. The upper pictures of Figure 5.3 suggest that, as is the case with canonical Hamiltonian systems, k may play a beneficial role also during the simulation of non-canonical Hamiltonian problems.

In fact, from the upper-right picture of Figure 5.3, we may observe that $|\tilde{p}|$ (see (5.57)) remains quite small during the simulation. Consistently with the Remark on page 163, even though the numerical solution computed by the HBVM(k, 1) method is only second order accurate in time, we expect the error in the Hamiltonian (5.49) to decrease with order $2k$, until it reaches the same order of magnitude of the error on the $\tilde{p}$ component, in accord with (5.59).

The pictures presented in Figure 5.4 display the solution error (solid line), the error in the Hamiltonian H (dashed line), and the error on $\tilde{p}$ (dotted line with circles), at time $t = 100$, plotted versus the time-step h, for $k = 1, 2, 3, 4$. As one may see, the solution error always decreases with order 2, as well as the error on $\tilde{p}$ (which is, however, quite small, $\simeq 10^{-12} \equiv \varepsilon$). On the other hand, the error on the Hamiltonian approximately decreases with order $2k$, as long as it remains larger than the error on $\tilde{p}$. This is indeed the case for $k = 1, 2, 3$. For $k = 4$, the Hamiltonian error displays an initial decrease rate of order 8 for larger stepsizes: as h is further reduced, it becomes comparable with the error on $\tilde{p}$ and, thereafter, it starts decreasing approximately with order 2.

To conclude this section, even though we shall not tackle the problem of the efficient implementation of HBVMs for problems in the form (5.1), we compare them with some known explicit methods. In particular, we solve problem (5.60)-(5.61), with $\gamma = 1$ and periodic boundary conditions, on the time interval $[0, 100]$, by means of the following methods:[8]

order 2: the HBVM(5, 1) method, and the symplectic Störmer-Verlet method (SV-2);

order 4: the HBVM(6, 2) method, and the composition method (SV-4) based on the symplectic Störmer-Verlet method (each step requiring 3 steps of the basic method);

---

[8]For the composite Störmer-Verlet methods, we refer to page 22.

TABLE 5.1: Parameters used for constructing Figure 5.5.

Method	$h_{\max}$	$h_{\min}$	ν
HBVM(5,1)	0.5	0.003	10
HBVM(6,2)	0.5	0.1	4
HBVM(9,3)	1	0.25	4
SV2	0.1	0.0006	13
SV4	0.1	0.007	7
SV6	0.1	0.01	5

order 6: the HBVM$(9,3)$ method, and the composition method (SV-6) based on the symplectic Störmer-Verlet method (each step requiring 9 steps of the basic method).

A Fourier-Galerkin space discretization with $N = 100$ and $m = 200$ (see (5.37) and (5.43), respectively) has been used to obtain the semi-discrete problem. To compare the methods, we construct a corresponding *Work-Precision Diagram*, by following the standard used in the *Test Set for IVP Solvers*.[9] In more detail, we plot the accuracy, measured in terms of the maximum absolute error, w.r.t. the execution time. The curve of each method is obtained by using ν (logarithmically) equispaced steps between $h_{\min}$ and $h_{\max}$, as specified in Table 5.1.[10] When the stepsize used does not exactly divide the final time $T = 100$, the nearest mesh-point is considered. Figure 5.5 summarizes the results, and one sees that the (practically) energy-conserving HBVMs are competitive, even with respect to explicit solvers of the same order.[11]

5.5 The nonlinear Schrödinger equation

We sketch the basic ideas to extend the analysis carried out for the semi-linear wave equation (5.1) to different Hamiltonian PDEs. In particular, we here consider the nonlinear Schrödinger equation (in dimensionless form),

$$\mathrm{i}\psi_t + \psi_{xx} + 2\kappa|\psi|^2\psi = 0, \qquad (x,t) \in (0,1) \times [0,\infty), \qquad \psi(x,0) = \phi_0(x), \tag{5.62}$$

where i denotes, as usual, the imaginary unit. Setting $\psi = u + \mathrm{i}v$, we obtain the real form of (5.62),

---

[9] https://www.dm.uniba.it/~testset/testsetivpsolvers/

[10] Larger values of $h_{\max}$ for the explicit methods (see Table 5.1) are not allowed because of stability issues.

[11] All tests have been carried out on a computer with a 2.8 GHz i7 2-core processor, running Matlab, rel. 2014b.

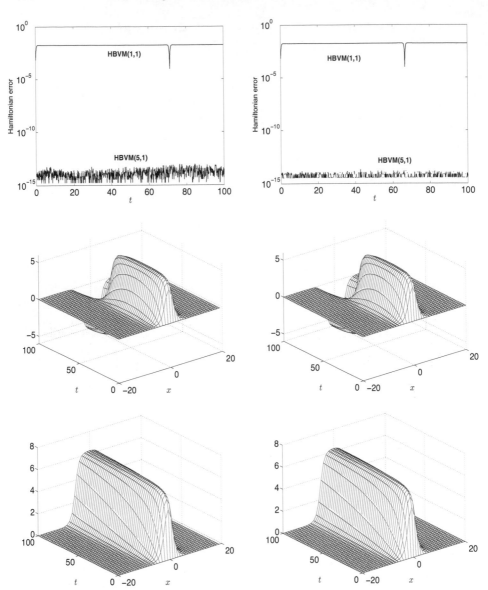

FIGURE 5.2: Sine-Gordon problem (5.60)-(5.61) with $\gamma = 1$ and periodic boundary conditions. On the left, from up to down, finite-difference space discretization: absolute error in the numerical Hamiltonian by using HBVM(1,1) and HBVM(5,1); solution computed by HBVM(1,1); solution computed by HBVM(5,1). On the right, from up to down, Fourier-Galerkin space discretization: absolute error in the numerical Hamiltonian by using HBVM(1,1) and HBVM(5,1); solution computed by HBVM(1,1); solution computed by HBVM(5,1). Time-step used: $h = 10^{-1}$.

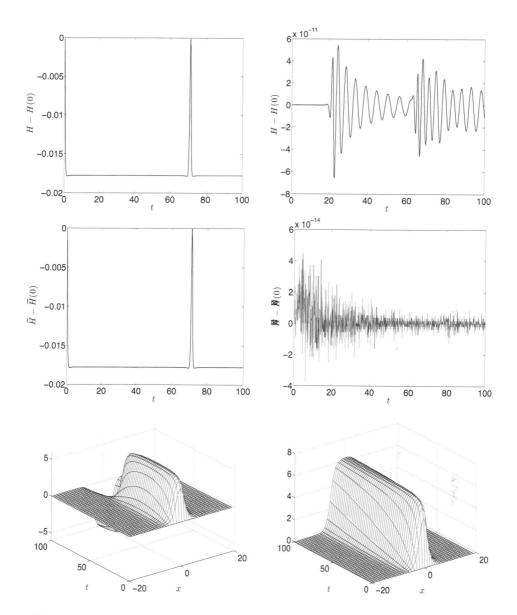

FIGURE 5.3: Sine-Gordon problem (5.60)-(5.61) with $\gamma = 1$ and Dirichlet boundary conditions. On the left, from up to down, results for HBVM(1,1): numerical value of $H - H(0)$ (see (5.49)); numerical value of $\widetilde{H} - \widetilde{H}(0)$ (see (5.55)); computed solution. On the right, from up to down, results for HBVM(5,1): numerical value of $H - H(0)$; numerical value of $\widetilde{H} - \widetilde{H}(0)$; computed solution. Time-step used: $h = 10^{-1}$.

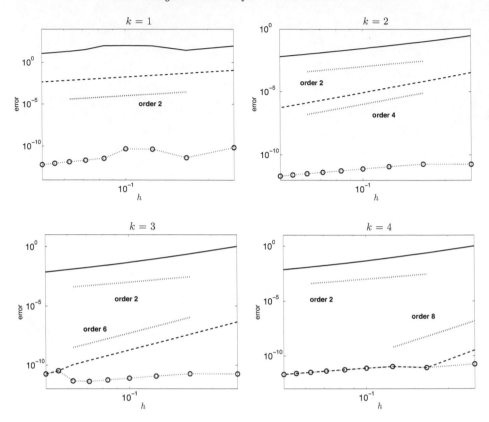

FIGURE 5.4: Sine-Gordon problem (5.60)-(5.61) with $\gamma = 1$, Dirichlet boundary conditions and finite-difference discretization with $N = 400$, solved by means of HBVM$(k,1)$, $k = 1, 2, 3, 4$. Errors versus time-step: solution (solid line), Hamiltonian function (dashed line), and $\tilde{p}$ (dotted line with circles) at $t = 100$.

$$
\begin{aligned}
u_t &= -v_{xx} - 2\kappa(u^2 + v^2)v, \qquad (x,t) \in (0,1) \times [0,\infty), \qquad (5.63) \\
v_t &= u_{xx} + 2\kappa(u^2 + v^2)u,
\end{aligned}
$$

which is Hamiltonian with Hamiltonian functional (compare with (5.7))

$$
\begin{aligned}
\mathcal{H}[u,v](t) &= \frac{1}{2} \int_0^1 \left[u_x^2(x,t) + v_x^2(x,t) - \kappa \left(u^2(x,t) + v^2(x,t) \right)^2 \right] \mathrm{d}x \\
&=: \int_0^1 E(x,t)\, \mathrm{d}x. \qquad (5.64)
\end{aligned}
$$

Again, (5.63) can be formally recast as in (5.8)–(5.10), with the new Hamiltonian function (5.64). In order to repeat for (5.63) the procedure leading to

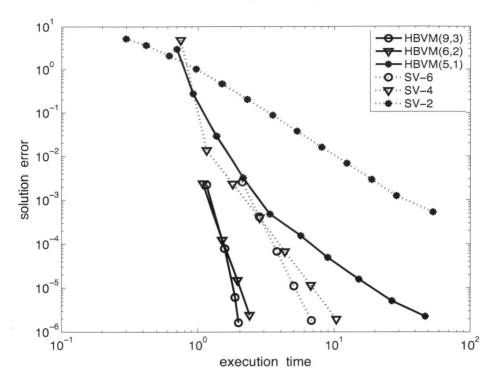

FIGURE 5.5: Sine-Gordon problem (5.60)-(5.61), $\gamma = 1$, periodic boundary conditions. Work-precision diagram (accuracy vs. execution time) comparing HBVMs with explicit methods of the same order.

the Hamiltonian semi-discretization of (5.1) with either periodic, or Dirichlet, or Neumann boundary conditions, it is enough to derive the conservation law corresponding to (5.13) and the expression of the Hamiltonian (5.15) (see page 149). Concerning the former conservation law, from (5.64) and (5.63) we have

$$E_t(x,t) = u_x(x,t)u_{xt}(x,t) + v_x(x,t)v_{xt}(x,t)$$

$$\underbrace{-2\kappa\left(u^2(x,t) + v^2(x,t)\right)u(x,t)}_{=\,(v_t - u_{xx})}u_t(x,t)\underbrace{-2\kappa\left(u^2(x,t) + v^2(x,t)\right)v(x,t)}_{=\,(u_t + v_{xx})}v_t(x,t)$$

$$= u_x(x,t)u_{xt}(x,t) + u_t(x,t)u_{xx}(x,t) + v_x(x,t)v_{xt}(x,t) + v_t(x,t)v_{xx}(x,t)$$

$$= \left(u_x(x,t)u_t(x,t)\right)_x + \left(v_x(x,t)v_t(x,t)\right)_x =: -F_x(x,t).$$

Consequently, in place of (5.13) we get

$$E_t(x,t) + F_x(x,t) = 0, \qquad F(x,t) = -u_x(x,t)u_t(x,t) - v_x(x,t)v_t(x,t).$$

Similarly, from (5.64) we obtain the analogue of (5.15):

$$
\begin{aligned}
\mathcal{H}[u,v](t) &= \frac{1}{2}\int_0^1 \left[u_x^2(x,t) + v_x^2(x,t) - \kappa\left(u^2(x,t) + v^2(x,t)\right)^2 \right] \mathrm{d}x \\
&= \frac{1}{2}\int_0^1 \Big[-u(x,t)u_{xx}(x,t) + (u(x,t)u_x(x,t))_x \\
&\qquad -v(x,t)v_{xx}(x,t) + (v(x,t)v_x(x,t))_x - \kappa\left(u^2(x,t) + v^2(x,t)\right)^2 \Big]\mathrm{d}x \\
&= -\frac{1}{2}\int_0^1 \left[u(x,t)u_{xx}(x,t) + v(x,t)v_{xx}(x,t) + \kappa\left(u^2(x,t) + v^2(x,t)\right)^2 \right] \mathrm{d}x + \\
&\qquad \frac{1}{2}\left[u(1,t)u_x(1,t) - u(0,t)u_x(0,t) + v(1,t)v_x(1,t) - v(0,t)v_x(0,t) \right].
\end{aligned}
$$

Starting from these relations, the space discretization can be carried out, by using either a finite-difference or a Fourier-Galerkin approach, as was done for (5.1).

Bibliographical notes

The numerical solution of Hamiltonian PDEs, computed by means of symplectic or multisyplectic methods, has been studied by several authors. To mention a few, we quote [20, 21, 22, 23, 74, 84, 91, 94, 114, 115, 126, 133, 134, 136, 138, 148, 172].

Conservation issues for Hamiltonian PDEs have been studied in [65, 76, 92, 121, 122, 123, 144, 146].

The use of methods conserving (a possibly discrete) energy is considered in, e.g., [67, 89, 95, 96, 125, 127, 135, 140, 141, 156, 157]. The approach used in this chapter is based on [29, 93].

Higher-order symmetric discretization formulae can be found in [10]. A thorough analysis of spectral methods can be found in the classical reference [66] (see also [128, 171]).

The soliton problem for the Sine-Gordon equation is taken from [173]. The numerical illustration of the Hamiltonian error, in the case of Dirichlet and Neumann boundary conditions, matches the analysis done in [129] for small non-conservative perturbations of conservative problems.

Chapter 6

Extensions

In this final chapter, we report a collage-style sketch of the ongoing research program related to the development of the theory of line integral methods. We exploit the discrete line integral tool to derive a few extensions of HBVMs for conserving multiple invariants in conservative systems. We also discuss the use of HBVMs to solve Hamiltonian boundary value problems. The extensions presented in this chapter are still under investigation and here we just report a mention to some preliminary results, hoping to convey the potentialities this new approach brings.

6.1 Conserving multiple invariants

Let us approach again the problem of approximating the solution of a Hamiltonian problem,

$$\dot{y} = J\nabla H(y) =: f(y), \qquad y(0) = y_0 \in \mathbb{R}^{2m}, \qquad t \in [0, h], \qquad (6.1)$$

by defining a suitable polynomial $\sigma(ch) \in \Pi_s$ such that

$$\sigma(0) = y_0, \qquad \sigma(h) =: y_1, \qquad H(y_1) = H(y_0),$$

thus yielding energy conservation. As a starting point for our discussion, we sketch once more the basic arguments underlying the definition of a Runge-Kutta line integral method described, in full detail, in Section 1.6 on page 32:

(i) We expand $\dot{\sigma}$ along the Legendre polynomial basis,

$$\dot{\sigma}(ch) = \sum_{j=0}^{s-1} P_j(c)\gamma_j, \qquad c \in [0, 1], \qquad (6.2)$$

where the coefficients γ_j are, for the moment, unknown.

(ii) Consequently, we obtain

$$\sigma(ch) = y_0 + h \sum_{j=0}^{s-1} \int_0^c P_j(x)\mathrm{d}x \, \gamma_j, \qquad c \in [0, 1], \qquad (6.3)$$

173

with the new approximation given by

$$y_1 := \sigma(h) = y_0 + h\gamma_0.$$

(iii) We evaluate the line integral of $\nabla H(y)$ along the curve $\sigma(ch)$ in the phase space,

$$
\begin{aligned}
H(y_1) - H(y_0) \;&=\; H(\sigma(h)) - H(\sigma(0)) \;=\; \int_0^h \nabla H(\sigma(t))^\top \dot\sigma(t)\mathrm{d}t \\
&=\; h\int_0^1 \nabla H(\sigma(ch))^\top \dot\sigma(ch)\mathrm{d}c \;=\; h\int_0^1 \nabla H(\sigma(ch))^\top \sum_{j=0}^{s-1} P_j(c)\gamma_j\mathrm{d}c \\
&=\; h\sum_{j=0}^{s-1} \left[\int_0^1 P_j(c)\nabla H(\sigma(ch))\mathrm{d}c\right]^\top \gamma_j.
\end{aligned}
$$

(iv) To get energy conservation, we impose the orthogonality conditions

$$\gamma_j \;:=\; J\int_0^1 P_j(c)\nabla H(\sigma(ch))\mathrm{d}c, \qquad j = 0,\dots,s-1. \qquad (6.4)$$

As was seen in Chapters 1 and 3, the procedure described above leads to the definition of an energy-conserving Runge-Kutta method (HBVM), upon approximating the integrals in (6.4) by means of an interpolatory quadrature formula. The polynomial $\sigma(ch)$ defined at (6.3), though assuring the energy conservation condition, fails, in general, to yield the conservation of further possible first integrals that a Hamiltonian system may posses. Considering that, given a first integral $L(y)$, the local error is

$$L(\sigma(h)) - L(y_0) = O(h^{2s+1}),$$

a natural approach to attack the question of conserving further first integrals is to slightly perturb the polynomial $\sigma(ch)$ by introducing a set of free parameters in its definition.

To this end, we replace (6.4) described at point (iv) above, with

$$\gamma_j = \eta_j J\int_0^1 P_j(c)\nabla H(\sigma(ch))\mathrm{d}c, \qquad j = 0,\dots,s-1, \qquad (6.5)$$

with η_j suitable scalars. The energy conservation condition is evidently retained by the new polynomial $\sigma(ch)$ in (6.3) defined according to (6.5) rather than (6.4).[1]

Choosing all $\eta_j = 1$ leads us back to the original formulae (6.4). On the other hand, allowing some η_j different from (but close to) one, will act as

---

[1]Since from now on we will make no use of the original definition (6.4), we continue to denote by $\sigma(ch)$ the polynomial at (6.3) defined through the perturbed coefficients (6.5).

introducing a perturbation in the polynomial curve $\sigma(ch)$ defined at (6.3), and we will take advantage of this gained freedom to impose the conservation of further possible invariants of the Hamiltonian system at hand. The following result, which generalizes that of Theorem 3.3 on page 87, tells us what the entity of perturbations of the scalar coefficients η_j should be, in order that the resulting formula may retain the same order $2s$ as the underlying unperturbed one (the HBVM method). In the sequel, we denote by $\zeta_j(\sigma)$ the coefficients of the Fourier expansion of $f(\sigma(ch))$ along the Legendre polynomial basis:

$$\zeta_j(\sigma) := \int_0^1 P_j(c)f(\sigma(ch))\mathrm{d}c, \qquad j = 0, 1, \ldots. \tag{6.6}$$

From (6.5), it follows that $\gamma_j = \eta_j\zeta_j(\sigma)$, $j = 0, \ldots, s-1$.

Corollary 6.1. *With reference to (6.1)–(6.5), and provided that*

$$\eta_j = 1 + \alpha_j(h), \qquad with \qquad \alpha_j(h) = O(h^{2(s-j)}), \qquad j = 0, \ldots, s-1, \tag{6.7}$$

one has[2]

$$y_1 - y(h) = O(h^{2s+1}).$$

Proof. Adopting the same notation as in Theorem 3.3, we obtain, by virtue of (6.7):

$$\begin{aligned}
y_1 - y(h) &= \sigma(h) - y(h) \\[4pt]
&= y(h; h, \sigma(h)) - y(h; 0, \sigma(0)) = \int_0^h \frac{\mathrm{d}}{\mathrm{d}t} y(h; t, \sigma(t))\,\mathrm{d}t \\[4pt]
&= \int_0^h \left[\frac{\partial}{\partial \tilde{t}} y(h; \tilde{t}, \sigma(t))\big|_{\tilde{t}=t} + \frac{\partial}{\partial \tilde{y}} y(h; t, \tilde{y})\big|_{\tilde{y}=\sigma(t)} \dot{\sigma}(t) \right] \mathrm{d}t \\[4pt]
&= \int_0^h \left[-\Phi(h; t, \sigma(t))f(\sigma(t)) + \Phi(h; t, \sigma(t))\dot{\sigma}(t) \right] \mathrm{d}t \\[4pt]
&= h\int_0^1 \left[-\Phi(h; ch, \sigma(ch))f(\sigma(ch)) + \Phi(h; ch, \sigma(ch))\dot{\sigma}(ch) \right] \mathrm{d}c \\[4pt]
&= h\int_0^1 \Phi(h; ch, \sigma(ch)) \left[\sum_{j=0}^{s-1} P_j(c)\gamma_j - \sum_{j \geqslant 0} P_j(c)\zeta_j(\sigma) \right] \mathrm{d}c \\[4pt]
&= h\int_0^1 \Phi(h; ch, \sigma(ch)) \left[\sum_{j=0}^{s-1} P_j(c)\zeta_j(\sigma)\alpha_j(h) - \sum_{j \geqslant s} P_j(c)\zeta_j(\sigma) \right] \mathrm{d}c
\end{aligned}$$

---

[2]That is, we still obtain an order $2s$ approximation to the solution, over any finite interval.

$$= h \sum_{j=0}^{s-1} \underbrace{\left[\int_0^1 P_j(c)\Phi(h; ch, \sigma(ch))dc \right] \zeta_j(\sigma)}_{O(h^{2j})} \overbrace{\alpha_j(h)}^{O(h^{2(s-j)})}$$

$$- h \sum_{j \geqslant s} \left[\int_0^1 P_j(c)\Phi(h; ch, \sigma(ch))dc \right] \zeta_j(\sigma) = O(h^{2s+1}). \quad \square$$

In the sequel, we shall actually exploit the possibility of choosing the co-efficients of the polynomial σ in the form (6.5)–(6.7), in order to impose the conservation of additional (functionally independent) first integrals, besides the Hamiltonian H. To this end, let us assume that (6.1) admits the following set of $\nu \geqslant 1$ invariants: $L : \mathbb{R}^{2m} \to \mathbb{R}^\nu$ such that

$$L(y(t)) \equiv L(y_0), \qquad \forall t \geqslant 0, \tag{6.8}$$

along the solution of (6.1) and for any admissible initial condition y_0 in a domain $\Omega \subset \mathbb{R}^{2m}$ which, for sake of simplicity, we shall assume to coincide with the whole space. Since we wish to tune the s free parameters η_j in order to get the conservation of all these ν invariants along the numerical solution, we assume that $s \geqslant \nu$ in the sequel. As we know, property (6.8) is equivalent to require that

$$\nabla L(y)^\top f(y) = 0, \qquad \forall y. \tag{6.9}$$

In addition to this, we shall assume that all the points belonging to the man-ifold

$$\mathcal{L}(y_0) = \{y \in \mathbb{R}^{2m} : L(y) = L(y_0)\} \tag{6.10}$$

are regular, so that $\operatorname{rank}(\nabla L(y)) = \nu$, $\forall y \in \mathcal{L}(y_0)$.[3] For any polynomial path σ one has that, by setting

$$\rho_j(\sigma) := \int_0^1 P_j(c)\nabla L(\sigma(ch))dc \in \mathbb{R}^{2m \times \nu}, \qquad j = 0, 1, \ldots, \tag{6.11}$$

the following expansions hold true (see (6.6)):[4]

$$\nabla L(\sigma(ch)) = \sum_{j \geqslant 0} P_j(c)\rho_j(\sigma), \qquad f(\sigma(ch)) = \sum_{j \geqslant 0} P_j(c)\zeta_j(\sigma), \qquad c \in [0,1]. \tag{6.12}$$

From (6.9) and the fact that the basis is orthonormal, we obtain

$$\mathbf{0} = \int_0^1 \nabla L(\sigma(ch))^\top f(\sigma(ch))dc = \sum_{j \geqslant 0} \rho_j(\sigma)^\top \zeta_j(\sigma).$$

[3] This is indeed the case, when the invariants are functionally independent of each other.
[4] As usual, assuming that both f and L are regular enough.

By virtue of Lemma 3.1 on page 86, this implies that

$$r_s(h) := \sum_{j=0}^{s-1} \rho_j(\sigma)^\top \zeta_j(\sigma) = - \sum_{j \geq s} \rho_j(\sigma)^\top \zeta_j(\sigma) \equiv O(h^{2s}). \qquad (6.13)$$

Because of the result of Corollary 6.1, let us then look for scalars η_j in the form

$$\eta_j = 1 - \alpha_j h^{2(s-j)}, \qquad j = 0, \ldots, s-1, \qquad (6.14)$$

where, to conform with the assumption of Corollary 6.1, the coefficients α_j should be bounded away from zero as $h \to 0$, and are to be selected so that $L(\sigma(h)) = L(\sigma(0)) \equiv L(y_0)$. In order to avoid the trivial solution

$$\alpha_j h^{2(s-j)} = 1, \quad j = 0, \ldots, s-1, \quad \Rightarrow \quad \eta_j = 0, \quad j = 0, \ldots, s-1,$$

during the practical numerical computation of the coefficients α_j, we impose the constraint that at least one of the α_j be zero, thus assuming now $\nu > s$. As an example, we can set

$$\alpha_0 = 0, \qquad \alpha_j = 0, \qquad j = \nu+1, \ldots, s-1, \qquad (6.15)$$

though different choices are possible. In so doing, we have that $\eta_0 = 1$ and $\eta_{\nu+1} = \cdots = \eta_{s-1} = 1$. Clearly, if the remaining scalars α_j are $O(1)$, the result in Corollary 6.1 continues to hold. By imposing the conservation of the ν invariants, one then obtains:

$$
\begin{aligned}
0 &= \int_0^1 \nabla L(\sigma(ch))^\top \dot{\sigma}(ch) dc = \int_0^1 \nabla L(\sigma(ch))^\top \sum_{j=0}^{s-1} P_j(c) \gamma_j dc \\
&= \sum_{j=0}^{s-1} \int_0^1 P_j(c) \nabla L(\sigma(ch))^\top dc \gamma_j = \sum_{j=0}^{s-1} \rho_j(\sigma)^\top \gamma_j \\
&= r_s(h) - \sum_{j=1}^{\nu} \rho_j(\sigma)^\top \zeta_j(\sigma) \alpha_j h^{2(s-j)},
\end{aligned}
$$

where $r_s(h)$ is the vector with $O(h^{2s})$ entries defined at (6.13). The previous set of equations can be cast in vector form as

$$\Psi_\nu(h)\alpha = r_s(h), \qquad (6.16)$$

where

$$
\begin{aligned}
\alpha &= \begin{pmatrix} \alpha_1 & \ldots & \alpha_\nu \end{pmatrix}^\top, \\
\Psi_\nu(h) &= \begin{pmatrix} h^{2(s-1)} \rho_1(\sigma)^\top \zeta_1(\sigma) & \ldots & h^{2(s-\nu)} \rho_\nu(\sigma)^\top \zeta_\nu(\sigma) \end{pmatrix} \in \mathbb{R}^{\nu \times \nu}.
\end{aligned}
$$

We observe that both the coefficient matrix $\Psi_\nu(h)$ and the right-hand side $r_s(h)$ in (6.16) have $O(h^{2s})$ entries so, assuming that the inverse of $\Psi_\nu(h)$ has

---

Algorithm 6.1 - EHBVM(k, s)

---

compute

$$u(ch) = y_0 + h \sum_{j=0}^{s-1} \int_0^c P_j(x)\mathrm{d}x\, \hat\zeta_j \left(1 - h^{2(s-j)}\hat\alpha_j\right), \quad c \in [0,1],$$

such that :

$$\hat\zeta_j = \sum_{\ell=1}^k b_\ell P_j(c_\ell) f(u(c_\ell h)),$$

$$\hat\rho_j = \sum_{\ell=1}^k b_\ell P_j(c_\ell) \nabla L(u(c_\ell h)), \qquad j = 0, \ldots, s-1,$$

$$\hat\Psi_s(h) = \left(h^{2(s-1)}\hat\rho_1^\top \hat\zeta_1 \quad \cdots \quad h^{2(s-\nu)}\hat\rho_\nu^\top \hat\zeta_\nu \right),$$

$$\hat r_s(h) = \sum_{j=0}^{s-1} \hat\rho_j^\top \hat\zeta_j,$$

$$\hat{\boldsymbol\alpha} := \left(\hat\alpha_1 \quad \cdots \quad \hat\alpha_\nu \right)^\top \equiv \hat\Psi_\nu(h)^{-1}\hat r_s(h),$$

$$\hat\alpha_0 = 0, \qquad \hat\alpha_j = 0, \quad j = \nu+1, \ldots, s-1,$$

set :

$$y_1 := u(h) \equiv y_0 + h\hat\zeta_0.$$

$O(h^{-2s})$ entries, we obtain $\boldsymbol\alpha = O(1)$. Consequently, by virtue of Corollary 6.1, the order of the approximation remains $2s$. We observe that the normalization $\alpha_0 = 0$ in (6.15) implies that the new approximation to $y(h)$ turns out to be given by

$$y_1 := \sigma(h) = y_0 + h\zeta_0(\sigma) \equiv y_0 + h \int_0^1 f(\sigma(ch))\mathrm{d}c. \qquad (6.17)$$

It is worth mentioning that the conservation of linear invariants (i.e., $L(y) = c^\top y$) needs not be imposed. In fact, by virtue of (6.9), they are implicitly conserved.

As usual, in order to obtain a numerical method, we must approximate the integrals in (6.6) and (6.11) by means of suitable quadrature rules. By choosing a Gaussian quadrature based at k points (c_i, b_i), one obtains, in place of the polynomial σ, a generally different polynomial $u \in \Pi_s$, which is defined by the Algorithm 6.1. In principle, one could use a different quadrature rule for each invariant (and for the Hamiltonian). In the sequel, for sake of simplicity, we shall always use the same quadrature for all of them.

Definition 6.1. *We name* Enhanced Hamiltonian Boundary Value Method *with k stages and degree s, in short* EHBVM(k, s), *the method defined by Algorithm 6.1.*

The following result is easily established, by using arguments similar to those used for proving Theorems 3.5 and 3.4 on pages 91-92.

Theorem 6.1. *Let $y_1 \approx y(h)$, solution of (6.1), be the approximation provided by the EHBVM(k, s) method defined by Algorithm 6.1. Then*

$$y_1 - y(h) = O(h^{2s+1}).$$

Moreover,

$$H(y_1) - H(y_0) = \begin{cases} 0, & \text{if } H \text{ is a polynomial of degree} \\ & \text{not larger than } 2k/s, \\ O(h^{2k+1}), & \text{otherwise,} \end{cases}$$

and, with reference to the invariants (6.8),

$$L(y_1) - L(y_0) = \begin{cases} 0, & \text{if } L \text{ is a polynomial of degree} \\ & \text{not larger than } 2k/s, \\ O(h^{2k+1}), & \text{otherwise.} \end{cases}$$

We observe that the EHBVM(k, s) method defined by Algorithm 6.1 admits a Runge-Kutta like representation, by regarding

$$Y_i := u(c_i h), \qquad i = 1, \ldots, k,$$

as the corresponding stages. In fact, using the usual matrices defined at (3.33)–(3.35) (see page 95), one obtains the following modified Butcher tableau:

$$\frac{c \mid \mathcal{I}_s \Lambda_s \mathcal{P}_s^\top \Omega}{\quad b^\top}, \qquad \Lambda_s = \begin{pmatrix} 1 & & & \\ & 1 - h^{2(s-1)}\alpha_1 & & \\ & & \ddots & \\ & & & 1 - h^2 \alpha_{s-1} \end{pmatrix},$$

where we recall that, according to (6.15), $\alpha_j = 0$, for $j > \nu$. We stress, however, that this is not actually a Runge-Kutta method, since matrix Λ_s does depend on the stage vector (thus enforcing the conservation of the invariants).

In order to give evidence of the previous results, let us consider the Kepler problem described in Section 2.5 on page 59. For such a problem, there are two additional (functionally independent) invariants, besides the Hamiltonian H (see (2.18)), i.e., the angular momentum M and the second component of the Lenz vector, A_2, defined at (2.29) and (2.30), respectively (see page 64, ff.). Since we have two additional invariants, we have to use a value of $s \geqslant 3$. Let us

TABLE 6.1: Numerical results for the Kepler problem.

HBVM$(3,3)$								
h	error	rate	$\|H - H_0\|$	rate	$\|M - M_0\|$	rate	$\|A_2 - A_{2,0}\|$	rate
$T/20$	5.4942e-02	–	9.8093e-04	–	1.9984e-15	–	2.7808e-03	–
$T/40$	1.2567e-02	2.13	2.1303e-05	5.52	1.4433e-15	–	2.4704e-04	3.49
$T/80$	1.6453e-04	6.26	2.7226e-07	6.29	7.7716e-16	–	4.7339e-06	5.71
$T/160$	2.4879e-06	6.05	4.1118e-09	6.05	3.3307e-15	–	7.5491e-08	5.97
$T/320$	3.8609e-08	6.01	6.3792e-11	6.01	3.8858e-15	–	1.1856e-09	5.99
$T/640$	6.0192e-10	6.00	9.9720e-13	6.00	2.1094e-15	–	1.8551e-11	6.00
HBVM$(12,3)$								
h	error	rate	$\|H - H_0\|$	rate	$\|M - M_0\|$	rate	$\|A_2 - A_{2,0}\|$	rate
$T/20$	1.1177e-02	–	3.3307e-16	–	8.4119e-05	–	1.2003e-02	–
$T/40$	2.4870e-04	5.49	2.2204e-16	–	1.2764e-06	6.04	2.7924e-04	5.43
$T/80$	4.1502e-06	5.91	4.4409e-16	–	1.9857e-08	6.01	4.7338e-06	5.88
$T/160$	6.5907e-08	5.98	4.4409e-16	–	3.1070e-10	6.00	7.5488e-08	5.97
$T/320$	1.0339e-09	5.99	4.4409e-16	–	4.8574e-12	6.00	1.1856e-09	5.99
$T/640$	1.6103e-11	6.00	6.6613e-16	–	7.6827e-14	5.98	1.8545e-11	6.00
EHBVM$(12,3)$								
h	error	rate	$\|H - H_0\|$	rate	$\|M - M_0\|$	rate	$\|A_2 - A_{2,0}\|$	rate
$T/20$	5.1632e-03	–	4.4409e-16	–	2.2204e-16	–	2.2204e-16	–
$T/40$	1.0709e-04	5.59	4.4409e-16	–	2.2204e-16	–	2.2204e-16	–
$T/80$	1.7287e-06	5.95	4.4409e-16	–	2.2204e-16	–	3.3307e-16	–
$T/160$	2.7214e-08	5.99	6.6613e-16	–	2.2204e-16	–	3.3307e-16	–
$T/320$	4.2631e-10	6.00	4.4409e-16	–	2.2204e-16	–	3.3307e-16	–
$T/640$	6.7260e-12	5.99	6.6613e-16	–	2.2204e-16	–	5.5511e-16	–

then set $s = 3$, so that the methods have order 6, and consider the HBVM$(3,3)$ (i.e., the Gauss method of order six), HBVM$(12,3)$ and EHBVM$(12,3)$ methods. For the first method we then expect only momentum conservation, for the second only energy conservation, whereas for the last one we expect an (at least practical) conservation of all the three invariants.

We choose the initial conditions (4.43), with an eccentricity $e = 0.6$, so that the solution is periodic with period $T = 2\pi$, and apply the above mentioned methods with stepsize $h = T/n$, for increasing values of n, to cover 5 periods. In Table 6.1 we list the obtained results concerning the errors in the solution and in the invariants:[5] as one may see, the latter errors (when not conserved) decrease with order 6 for HBVM$(3,3)$ and HBVM$(12,3)$, whereas they are all very small for EHBVM$(12,3)$. Moreover, it is worth noticing that the solution error of the last method is more favorable than the others.

---

[5]The initial values of H, M, and A_2 are denoted H_0, M_0, and $A_{2,0}$, respectively. Moreover, in the tables $\| \cdot \|$ will denote the maximum norm over the considered interval.

6.2 General conservative problems

Let us now consider the case where the given problem is not Hamiltonian but has, nonetheless, a set of functionally independent invariants. We are now dealing with a general problem in the form

$$\dot{y} = f(y), \qquad y(0) = y_0 \in \mathbb{R}^m, \tag{6.18}$$

such that (6.8)-(6.9) continue to hold with $L : \mathbb{R}^m \to \mathbb{R}^\nu$ and, moreover, the points of the manifold (6.10) are all regular for the constraints. Consequently, the expansions given by (6.6), (6.11), and (6.12) continue formally to hold in the present context (assuming, as usual, that f and L are suitably regular). We shall look for a polynomial approximation $\sigma \in \Pi_s$ such that

$$\sigma(0) = y_0, \qquad y(h) \approx \sigma(h) =: y_1, \qquad L(y_1) = L(y_0),$$

and still yielding a $O(h^{2s+1})$ approximation. For this purpose, let us choose $\sigma(ch)$ as in (6.2)-(6.3), with the coefficients γ_j in the following form:

$$\gamma_j = \zeta_j(\sigma) - \rho_j(\sigma)\alpha_j, \qquad j = 0, \ldots, s-1, \tag{6.19}$$

where $\zeta_j(\sigma)$ and $\rho_j(\sigma)$ are defined at (6.6) and (6.11), respectively, and where now $\alpha_j \in \mathbb{R}^\nu$, $j = 0, \ldots, s-1$. The following result then holds true, extending to this setting the one proved in Corollary 6.1, the proof being almost the same.

Corollary 6.2. *With reference to (6.2)-(6.3) and (6.18)-(6.19), and provided that*

$$\alpha_j = \alpha_j(h) = O(h^{2(s-j)}), \qquad j = 0, \ldots, s-1,$$

one has $y_1 - y(h) = O(h^{2s+1})$.

Considering that each vector coefficient α_j brings ν free parameters, the easiest choice is to set all the α_j but one equal to zero. In the sequel, assuming $s \geqslant 2$, we will let α_1 the only coefficient in (6.19) possibly different from zero. The new polynomial $\sigma(ch)$ that will be used to advance the solution is then defined as

$$\dot{\sigma}(ch) = \sum_{j=0}^{s-1} P_j(c)\zeta_j(\sigma) - P_1(c)\rho_1(\sigma)\alpha_1(h), \qquad c \in [0,1], \tag{6.20}$$

and may be thought of as a perturbation of the polynomial defined at (6.2), where the perturbation term concerns its second coefficient.[6]

---

[6]In principle, any coefficient could be perturbed. In particular, when $s = 1$ one pertubs the first one, even though, in such a case, the conservation of possible linear invariants could be lost (unless explicitly imposed).

Let us see how the vector $\alpha_1(h)$ may be tuned in order to obtain the conservation of all invariants, namely $L(y_1) = L(y_0)$, $y_1 := \sigma(h)$ being the approximation to $y(h)$, with $\sigma(0) = y_0$. As usual, the vector $\alpha_1(h)$ will be derived through a line integral:

$$
\begin{aligned}
\mathbf{0} &= L(y_1) - L(y_0) = L(\sigma(h)) - L(\sigma(0)) \\
&= \int_0^h \nabla L(\sigma(t))^\top \dot\sigma(t) \mathrm{d}t = h \int_0^1 \nabla L(\sigma(ch))^\top \dot\sigma(ch) \mathrm{d}c \\
&= h \int_0^1 \nabla L(\sigma(ch))^\top \left[\sum_{j=0}^{s-1} P_j(c)\zeta_j(\sigma) - P_1(c)\rho_1(\sigma)\alpha_1(h) \right] \mathrm{d}c \\
&= h \left[\sum_{j=0}^{s-1} \rho_j(\sigma)^\top \zeta_j(\sigma) - \rho_1(\sigma)^\top \rho_1(\sigma)\alpha_1(h) \right].
\end{aligned}
$$

Consequently, one has that the conservation of the invariants will follow from the following requirement (see (6.13)):

$$
\left[\rho_1(\sigma)^\top \rho_1(\sigma) \right] \alpha_1(h) = \sum_{j=0}^{s-1} \rho_j(\sigma)^\top \zeta_j(\sigma) \equiv r_s(h) = O(h^{2s}).
$$

Since $\rho_1(\sigma)^\top \rho_1(\sigma)$ is symmetric and positive definite, with $O(h^2)$ entries, one obtains that $\alpha_1(h) = O(h^{2(s-1)})$. Consequently, the result of Corollary 6.2 applies and, therefore:

$$
L(y_1) = L(y_0), \qquad y_1 - y(h) = O(h^{2s+1}).
$$

It is worth mentioning that, since $\alpha_0(h) = 0$ in (6.20), also in this case the solution is advanced as in (6.17).

Again, in order to obtain a numerical method, we must approximate the integrals in (6.6) and (6.11) by means of suitable quadrature rules. Employing a Gaussian quadrature formula based at k points (c_i, b_i), one obtains, in place of the polynomial $\sigma(ch)$, a generally different polynomial $u(ch) \in \Pi_s$, which is defined by the Algorithm 6.2.[7]

Definition 6.2. *We name* Generalized Hamiltonian Boundary Value Method *with* k *stages and degree* s, *in short* GHBVM(k, s), *the method defined by Algorithm 6.2.*

The following result, analogous to Theorem 6.1 for EHBVMs, is easily established.

Theorem 6.2. *Let* $y_1 \approx y(h)$, *solution of (6.18), be the approximation provided by the GHBVM(k, s) method defined by Algorithm 6.2. Then*

$$
y_1 - y(h) = O(h^{2s+1}).
$$

---

[7] Also in the present case, one could use different quadrature rules for each invariant.

Algorithm 6.2 - GHBVM(k, s)

---

compute

$$u(ch) \;=\; y_0 + h\left[\sum_{j=0}^{s-1}\int_0^c P_j(x)\mathrm{d}x\hat{\zeta}_j \;-\; \int_0^c P_1(x)\mathrm{d}x\hat{\rho}_1\hat{\alpha}_1\right], \quad c \in [0,1],$$

such that :

$$\hat{\zeta}_j \;=\; \sum_{\ell=1}^k b_\ell P_j(c_\ell)f(u(c_\ell h)),$$

$$\hat{\rho}_j \;=\; \sum_{\ell=1}^k b_\ell P_j(c_\ell)\nabla L(u(c_\ell h)), \qquad j = 0,\ldots,s-1,$$

$$\hat{\alpha}_1 \;=\; [\hat{\rho}_1^\top\hat{\rho}_1]^{-1}\sum_{j=0}^{s-1}\hat{\rho}_j^\top\hat{\zeta}_j,$$

set :

$$y_1 \;:=\; u(h) \;\equiv\; y_0 + h\hat{\zeta}_0.$$

Moreover, concerning the conservation of the invariants, one has:

$$L(y_1) - L(y_0) = \begin{cases} 0, & \text{if } L \text{ is a polynomial of degree} \\ & \quad \text{not larger than } 2k/s, \\ O(h^{2k+1}), & \text{otherwise.} \end{cases}$$

As is the case with EHBVMs, Algorithm 6.2 is appropriate for methods of order at least four ($s \geqslant 2$). If $s = 1$, one should perturb the first coefficient, $\hat{\zeta}_0$, of the polynomial u.

It is also worth mentioning that the GHBVMs defined by Algorithm 6.2 can be also applied to Hamiltonian problems. In such a case, however, the conservation of the Hamiltonian must be explicitly imposed among the invariants brought by the function $L(y)$. As an example, let us consider again the Kepler problem described in Section 2.5, with initial conditions (4.43) and an eccentricity $e = 0.6$, so that the solution is periodic with period $T = 2\pi$. We consider the GHBVM$(8,2)$ method with stepsize $h = T/n$, for increasing values of n, to cover 5 periods. In Table 6.2 we list the results concerning the errors in the solution and in the invariants. As one may see, the method is fourth-order, according to Theorem 6.2, whereas the error in the invariants (the Hamiltonian H, the angular momentum M, and the second component of Lenz vector, A_2) is negligible, excluding when the coarsest stepsize is used, where it is of the prescribed order of magnitude (observe that both H and A_2 are non-polynomial functions).

TABLE 6.2: Numerical results for the Kepler problem.

			GHBVM$(8,2)$		
h	error	rate	$\|H - H_0\|$	$\|M - M_0\|$	$\|A_2 - A_{2,0}\|$
$T/20$	3.5950e-02	–	7.3600e-09	4.6629e-15	8.5285e-09
$T/40$	2.6601e-03	3.76	6.6613e-16	2.2204e-16	3.3307e-16
$T/80$	1.7270e-04	3.95	4.4409e-16	2.2204e-16	3.3307e-16
$T/160$	1.0895e-05	3.99	6.6613e-16	2.2204e-16	4.4409e-16
$T/320$	6.8252e-07	4.00	4.4409e-16	2.2204e-16	3.3307e-16
$T/640$	4.2682e-08	4.00	6.6613e-16	2.2204e-16	4.4409e-16

As a further application, we consider the class of *Poisson problems*. A Poisson system takes the form

$$\dot{y} = B(y)\nabla H(y), \qquad y(0) \in \mathbb{R}^m, \qquad \text{with} \qquad B(y)^\top = -B(y), \qquad (6.21)$$

and thus may be interpreted as a generalization of a Hamiltonian system. For problems in the form (6.21), the Hamiltonian $H(y)$ is still a constant of motion, as well as any *Casimir function* $C(y)$, such that $\nabla C(y)^\top B(y) = 0$.

Sometimes, in order to obtain a more accurate numerical solution, it is important to be able to preserve both the Hamiltonian and the Casimirs. To elucidate this aspect, we consider the Lotka-Volterra problem of dimension 3, which is in the form (6.21) with skew-symmetric matrix

$$B(y) = \begin{pmatrix} 0 & c\,y_1 y_2 & bc\,y_1 y_3 \\ -c\,y_1 y_2 & 0 & -y_2 y_3 \\ -bc\,y_1 y_2 & y_2 y_3 & 0 \end{pmatrix}, \qquad (6.22)$$

and Hamiltonian function

$$H(y) = ab\,y_1 + y_2 - a\,y_3 + \mu_1 \ln y_2 - \mu_2 \ln y_3, \qquad (6.23)$$

with the parameters a, b, c, saisfying $abc = -1$. In particular, we consider the following parameters:

$$a = -2, \qquad b = -1, \qquad c = -0.5, \qquad \mu_1 = 1, \qquad \mu_2 = 2, \qquad (6.24)$$

and the initial point

$$y(0) = \begin{pmatrix} 1, & 1.9, & 0.5 \end{pmatrix}^\top \qquad (6.25)$$

which originates a periodic solution of period $T \approx 2.878130103817$. This problem admits also the following Casimir:

$$C(y) = ab \ln y_1 - b \ln y_2 + \ln y_3, \qquad (6.26)$$

besides the Hamiltonian. We apply the GHBVM$(8,2)$ method with a stepsize $h = T/30 \approx 0.1$, to solve problem (6.21)–(6.26) under the following three requirements:

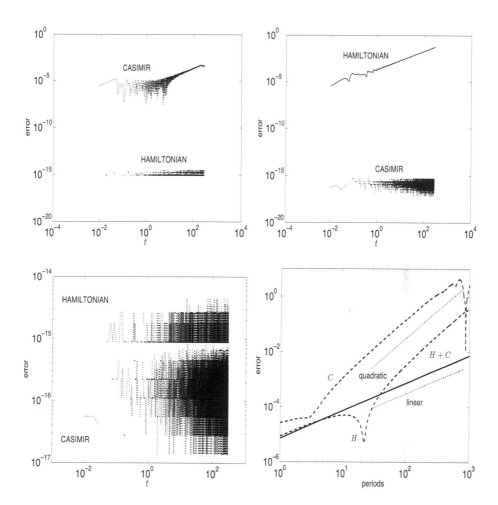

FIGURE 6.1: Numerical solution of problem (6.21)–(6.26) computed by the GHBVM(8, 2) method with stepsize $h \approx 10^{-1}$. Upper pictures: errors in the invariants when conserving only either the Hamiltonian (left) or the Casimir (right). Bottom-left picture: conservation of both the invariants to within machine precision. Bottom-right picture: error in the numerical solution when preserving only the Casimir (C, dashed line), only the Hamiltonian (H, dashed line), or both ($H + C$, solid line). To guide the eye, a typical linear and a quadratic growth are also shown (dotted lines).

1. conserve the Hamiltonian function H;

2. conserve the Casimir C;

3. conserve both H and C.

The upper pictures of Figure 6.1 show that the errors in the Casimir and in the Hamiltonian follow a linear growth, when their conservation is not required (cases 1 and 2, respectively). This fact clearly affects the error growth in the numerical solution, which turns out to be quadratic, when only one invariant is conserved (see the bottom-right picture of Figure 6.1). On the other hand, if we impose the concomitant conservation of both invariants (case 3), the error in the numerical solution follows a linear growth and, thus, turns out to be much more accurate in the long run.

6.3   EQUIP methods

The methods that we are going to study in this section stem from the symplecticity property of Gauss-Legendre Runge-Kutta methods, when applied to the canonical Hamiltonian problem (6.1). We recall that the Butcher tableau of the s-stage Gauss method may be cast as (see Section 1.4.3):

$$\frac{\boldsymbol{c} \quad A := \mathcal{P}_s X_s \mathcal{P}_s^\top \Omega}{\boldsymbol{b}^\top} \tag{6.27}$$

where

$$\mathcal{P}_s = \begin{pmatrix} P_0(c_1) & \cdots & P_{s-1}(c_1) \\ \vdots & & \vdots \\ P_0(c_s) & \cdots & P_{s-1}(c_s) \end{pmatrix}, \qquad \Omega = \begin{pmatrix} b_1 & & \\ & \ddots & \\ & & b_s \end{pmatrix},$$

and

$$X_s = \begin{pmatrix} \frac{1}{2} & -\xi_1 & & \\ \xi_1 & 0 & \ddots & \\ & \ddots & \ddots & -\xi_{s-1} \\ & & \xi_{s-1} & 0 \end{pmatrix}, \qquad \xi_i = \frac{1}{2\sqrt{4i^2 - 1}}, \qquad i = 1, \ldots, s-1,$$

are $s \times s$ matrices, and

$$\boldsymbol{b} = \begin{pmatrix} b_1, & \ldots, & b_s \end{pmatrix}^\top, \qquad \boldsymbol{c} = \begin{pmatrix} c_1, & \ldots, & c_s \end{pmatrix}^\top,$$

with (c_i, b_i) the nodes and weights of the Gauss-Legendre quadrature of order $2s$ (i.e., $P_s(c_i) = 0$, $i = 1, \ldots, s$.). Since $\mathcal{P}_s^\top \Omega = \mathcal{P}_s^{-1}$, (6.27) amounts to the

W-transformation of the method. The criterion for symplecticity of Runge-Kutta methods reported in Theorem 1.1 on page 19, namely

$$\Omega A + A^\top \Omega - \boldsymbol{b}\boldsymbol{b}^\top = O, \tag{6.28}$$

may be exploited to show the symplecticity property of formulae (6.27):

$$
\begin{aligned}
\Omega A + A^\top \Omega &= \Omega \mathcal{P}_s X_s \mathcal{P}_s^\top \Omega + \Omega \mathcal{P}_s X_s^\top \mathcal{P}_s^\top \Omega \\
&= \Omega \mathcal{P}_s \left(X_s + X_s^\top \right) \mathcal{P}_s^\top \Omega = \Omega \mathcal{P}_s \boldsymbol{e}_1 \boldsymbol{e}_1^\top \mathcal{P}_s^\top \Omega \\
&= \Omega \boldsymbol{1}\boldsymbol{1}^\top \Omega = \boldsymbol{b}\boldsymbol{b}^\top.
\end{aligned}
$$

Clearly, (6.28) will continue to hold, if we replace X_s with the matrix

$$X_s(\alpha) := X_s - \alpha W_s, \qquad \text{where} \qquad W_s^\top = -W_s. \tag{6.29}$$

For our purposes, we choose the skew-symmetric matrix W_s in the form:

$$W_s = \boldsymbol{e}_2 \boldsymbol{e}_1^\top - \boldsymbol{e}_1 \boldsymbol{e}_2^\top, \tag{6.30}$$

even though different choices are, in principle, allowed. The idea is now to select the parameter α in order to enforce energy conservation. In so doing, one obtains a parametric method that, at each time step, is defined by a symplectic map, so conserves all quadratic invariants and, at the same time, yields energy conservation. For this reason, these methods are named *Energy and QUadratic Invariants Preserving (EQUIP) methods*.[8]

These methods admit a Runge-Kutta like formulation, given by

$$\begin{array}{c|c} \boldsymbol{c} & \mathcal{P}_s X_s(\alpha) \mathcal{P}_s^\top \Omega \\ \hline & \boldsymbol{b}^\top \end{array} \tag{6.31}$$

with $X_s(\alpha)$ defined at (6.29)-(6.30). To define a strategy for the selection of the parameter α at each step of the integration procedure, we will resort again to the line integral approach. To this end, we need to determine the shape of the curve approximating $y(ch)$ in the time interval $[0, h]$, underlying formula (6.31). In this specific case, we will conveniently represent such a curve as the concatenation of two different polynomial curves in the phase space, $u, w \in \Pi_s$ that will be defined in the sequel.

We begin with observing that the Butcher matrix in (6.31) can be written as

$$
\begin{aligned}
\mathcal{P}_s \left(X_s - \alpha W_s \right) \mathcal{P}_s^\top \Omega &= \mathcal{P}_s X_s \left(I_s - \alpha X_s^{-1} W_s \right) \mathcal{P}_s^\top \Omega \\
&= \mathcal{I}_s \left[I_s - \alpha (\boldsymbol{\phi}_2 \boldsymbol{e}_1^\top - \boldsymbol{\phi}_1 \boldsymbol{e}_2^\top) \right] \mathcal{P}_s^\top \Omega,
\end{aligned}
$$

where

$$X_s^{-1} \boldsymbol{e}_i =: \boldsymbol{\phi}_i \equiv \begin{pmatrix} \phi_{i0} \\ \vdots \\ \phi_{i,s-1} \end{pmatrix}, \qquad i = 1, 2, \tag{6.32}$$

---

[8]Due to the dependence on the parameter α, such methods do not violate the non existence results presented in Chapter 1 (see page 19).

and we have taken into account that, since c contains the zeros of P_s,

$$
\mathcal{P}_s X_s = \mathcal{I}_s = \begin{pmatrix} \int_0^{c_1} P_0(x)dx & \cdots & \int_0^{c_1} P_{s-1}(x)dx \\ \vdots & & \vdots \\ \int_0^{c_s} P_0(x)dx & \cdots & \int_0^{c_s} P_{s-1}(x)dx \end{pmatrix}.
$$

(see (1.55)-(1.56) on page 25). Consequently, we get

$$
\begin{aligned}
Y &= \mathbf{1} \otimes y_0 + h\mathcal{I}_s \left[I_s - \alpha(\boldsymbol{\phi}_2 e_1^\top - \boldsymbol{\phi}_1 e_2^\top) \right] \mathcal{P}_s^\top \Omega \otimes I_{2m} f(Y) \\
&= \mathbf{1} \otimes y_0 + h\mathcal{I}_s \left[I_s - \alpha(\boldsymbol{\phi}_2 e_1^\top - \boldsymbol{\phi}_1 e_2^\top) \right] \otimes I_{2m}\hat{\gamma},
\end{aligned}
$$

having set

$$
\mathcal{P}_s^\top \Omega \otimes I_{2m} f(Y) =: \hat{\gamma} \equiv \begin{pmatrix} \hat{\gamma}_0 \\ \vdots \\ \hat{\gamma}_{s-1} \end{pmatrix}.
$$

Consistently with the approach used to derive HBVMs and all the methods in the present chapter, we now consider the polynomial $u(ch) \in \Pi_s$ satisfying the interpolation conditions: $u(0) = y_0$ and $u(ch) = Y$. It reads

$$
u(ch) = y_0 + h \sum_{j=0}^{s-1} \int_0^c P_j(x)dx \left[\hat{\gamma}_j - \alpha \left(\phi_{2j}\hat{\gamma}_0 - \phi_{1j}\hat{\gamma}_1 \right) \right], \quad c \in [0,1], \quad (6.33)
$$

and, therefore,

$$
\dot{u}(ch) = \sum_{j=0}^{s-1} P_j(c) \left[\hat{\gamma}_j - \alpha \left(\phi_{2j}\hat{\gamma}_0 - \phi_{1j}\hat{\gamma}_1 \right) \right], \quad c \in [0,1]. \quad (6.34)
$$

Unfortunately, the polynomial $u(ch)$ is not the right candidate to represent the correct action of method (6.31) since, on the one hand we have

$$
u(h) = y_0 + h \left[\hat{\gamma}_0 - \alpha \left(\phi_{20}\hat{\gamma}_0 - \phi_{10}\hat{\gamma}_1 \right) \right]
$$

while, from (6.31), we see that the new approximation is actually given by

$$
y_1 := y_0 + h\hat{\gamma}_0.
$$

Nonetheless, subtracting side by side the two latter centered formulae, we get

$$
y_1 - u(h) = h\alpha \left(\phi_{20}\hat{\gamma}_0 - \phi_{10}\hat{\gamma}_1 \right), \quad (6.35)
$$

so we can define a piecewise polynomial path in the phase space joining the initial state vector y_0 to the final state vector y_1, with the aid of the following additional polynomial:

$$
w(\tau) = \tau y_1 + (1-\tau)u(h) \quad \Rightarrow \quad \dot{w}(\tau) = y_1 - u(h) \equiv h\alpha \left(\phi_{20}\hat{\gamma}_0 - \phi_{10}\hat{\gamma}_1 \right).
$$

We are now in the right position to impose the energy-conservation property on method (6.31), by requiring that the line integral corresponding to the continuous path defined by the two polynomial curves $u(ch)$ and $w(\tau)$ does vanish:

$$
\begin{aligned}
0 \;=\; & H(y_1) - H(y_0) \;=\; H(w(1)) - H(w(0)) + H(u(h)) - H(u(0)) \\
=\; & \int_0^1 \nabla H(w(\tau))^\top \dot{w}(\tau)\mathrm{d}\tau \;+\; h\int_0^1 \nabla H(u(ch))^\top \dot{u}(ch)\mathrm{d}c \\
=\; & h\alpha \int_0^1 \nabla H(w(\tau))^\top \mathrm{d}\tau \,(\phi_{20}\hat{\gamma}_0 - \phi_{10}\hat{\gamma}_1) \\
& + h\int_0^1 \nabla H(u(ch))^\top \sum_{j=0}^{s-1} P_j(c)\left[\hat{\gamma}_j - \alpha\left(\phi_{2j}\hat{\gamma}_0 - \phi_{1j}\hat{\gamma}_1\right)\right]\mathrm{d}c \\
=\; & h\alpha \int_0^1 \nabla H(w(\tau))^\top \mathrm{d}\tau \,(\phi_{20}\hat{\gamma}_0 - \phi_{10}\hat{\gamma}_1) \\
& + h\sum_{j=0}^{s-1} \int_0^1 \nabla H(u(ch))^\top P_j(c)\mathrm{d}c\left[\hat{\gamma}_j - \alpha\left(\phi_{2j}\hat{\gamma}_0 - \phi_{1j}\hat{\gamma}_1\right)\right] \\
=\; & h\alpha\rho_0(w)^\top J\,(\phi_{20}\hat{\gamma}_0 - \phi_{10}\hat{\gamma}_1) \;+\; h\sum_{j=0}^{s-1} \gamma_j(u)^\top J\left[\hat{\gamma}_j - \alpha\left(\phi_{2j}\hat{\gamma}_0 - \phi_{1j}\hat{\gamma}_1\right)\right],
\end{aligned}
$$

where we have set

$$
\rho_0(w) \;:=\; J\int_0^1 \nabla H(w(\tau))\mathrm{d}\tau, \tag{6.36}
$$

$$
\gamma_j(u) \;:=\; J\int_0^1 \nabla H(u(ch))P_j(c)\mathrm{d}c, \qquad j = 0,\dots,s-1.
$$

Consequently, energy conservation is gained provided that

$$
\begin{aligned}
\alpha \;=\; & \frac{\sum_{j=0}^{s-1} \gamma_j(u)^\top J\hat{\gamma}_j}{-\rho_0(w)^\top J\,(\phi_{20}\hat{\gamma}_0 - \phi_{10}\hat{\gamma}_1) + \sum_{j=0}^{s-1} \gamma_j(u)^\top J\,(\phi_{2j}\hat{\gamma}_0 - \phi_{1j}\hat{\gamma}_1)} \\
\equiv\; & \frac{\sum_{j=0}^{s-1} \gamma_j(u)^\top J\hat{\gamma}_j}{(\gamma_0(u) - \rho_0(w))^\top J\,(\phi_{20}\hat{\gamma}_0 - \phi_{10}\hat{\gamma}_1) + O(h)},
\end{aligned} \tag{6.37}
$$

where the last equality follows from the fact that $\hat{\gamma}_j = O(h^j)$, $\gamma_j(u) = O(h^j)$, and $\rho_0(w) = O(1)$.[9]

The polynomial $u(ch)$ turns out to be also helpful in stating the order of the EQUIP methods. In fact, the following result holds true.

Theorem 6.3. *Let $u(h)$ be the approximation to $y(h)$, solution of (6.1), as defined by (6.33)–(6.37). The following properties hold true:*

---

[9] We observe that α implicitly appears at the right-hand sides in (6.37), since the polynomial u does depend on α.

(a) $u(h) - y(h) = O(h^{2s+1})$;

(b) $y_1 - y(h) = O(h^{2s+1})$.

Proof. In fact, by considering that $J^\top = -J$ and

$$\gamma_j(u) = O(h^j), \qquad \hat{\gamma}_j = \gamma_j(u) + O(h^{2s-j}), \qquad j = 0, \ldots, s-1, \qquad (6.38)$$

one obtains that the numerator in (6.37) is

$$\sum_{j=0}^{s-1} \gamma_j(u)^\top J \hat{\gamma}_j = \underbrace{\sum_{j=0}^{s-1} \gamma_j(u)^\top J \gamma_j(u)}_{=0} + O(h^{2s}) = O(h^{2s}).$$

Concerning the denominator in the same expression, one has that $\rho_0(w) = O(1)$. Moreover, it is possible to check that

$$(\phi_{10}, \phi_{20}) = \begin{cases} (2,0), & \text{when} \quad s \text{ is odd}, \\ (0, \xi_1^{-1}), & \text{when} \quad s \text{ is even}. \end{cases} \qquad (6.39)$$

Consequently, the denominator is $O(h)$ or $O(1)$, depending on the fact that s is odd or even, respectively. One then concludes that

$$\alpha = \begin{cases} O(h^{2s-1}), & \text{when} \quad s \text{ is odd}, \\ O(h^{2s}), & \text{when} \quad s \text{ is even}. \end{cases} \qquad (6.40)$$

In any event, by virtue of (6.37) and (6.39)-(6.40), one obtains that

$$\alpha\, (\phi_{2j}\hat{\gamma}_0 - \phi_{1j}\hat{\gamma}_1) = \begin{cases} O(h^{2s}), & j = 0, \\ O(h^{2s-1}), & j = 1, \ldots, s-1. \end{cases} \qquad (6.41)$$

Consequently, by taking into account (6.34) and (6.38), one has that the polynomial $\dot{u}(ch)$ can be written, for all $c \in [0,1]$, as:

$$\dot{u}(ch) = \sum_{j=0}^{s-1} P_j(c) [\hat{\gamma}_j(u) - \alpha\, (\phi_{2j}\hat{\gamma}_0 - \phi_{1j}\hat{\gamma}_1)] = \sum_{j=0}^{s-1} P_j(c) [\gamma_j(u) +$$

$$O(h^{2s-j}) - \alpha\, (\phi_{2j}\hat{\gamma}_0 - \phi_{1j}\hat{\gamma}_1)] = \sum_{j=0}^{s-1} P_j(c) [\gamma_j(u) + O(h^{2s-j})].$$

Therefore, property (a) follows by using an analogous argument exploited for proving Theorem 3.5 (see page 92). Property (b) is an immediate consequence of (6.35) and (6.41). $\qquad \square$

As usual, in order to obtain a numerical method, one needs to approximate the integrals in (6.36) by means of a quadrature formula. In so doing, the

---

Algorithm 6.3 - EQUIP(k, s)

compute

$$\bar{u}(ch) \;=\; y_0 + h \sum_{j=0}^{s-1} \int_0^c P_j(x)\mathrm{d}x \left[\hat{\gamma}_j - \alpha(\phi_{2j}\hat{\gamma}_0 - \phi_{1j}\hat{\gamma}_1) \right], \quad c \in [0,1],$$

such that :

$$\hat{\gamma}_j \;=\; \sum_{\ell=1}^{s} b_\ell P_j(c_\ell) J\nabla H(\bar{u}(c_\ell h)),$$

$$\bar{\gamma}_j \;=\; \sum_{\ell=1}^{k} \bar{b}_\ell P_j(\bar{c}_\ell) J\nabla H(\bar{u}(\bar{c}_\ell h)), \qquad j = 0, \ldots, s-1,$$

$$\bar{\rho}_0 \;=\; \sum_{\ell=1}^{k} \bar{b}_\ell J\nabla H(\bar{w}(\bar{c}_\ell)),$$

$$\alpha \;=\; \frac{\sum_{j=0}^{s-1} \bar{\gamma}_j^\top J\hat{\gamma}_j}{\sum_{j=0}^{s-1} \bar{\gamma}_j^\top J\left(\phi_{2j}\hat{\gamma}_0 - \phi_{1j}\hat{\gamma}_1\right) - \bar{\rho}_0^\top J\left(\phi_{20}\hat{\gamma}_0 - \phi_{10}\hat{\gamma}_1\right)},$$

set :

$$y_1 \;:=\; y_0 + h\hat{\gamma}_0.$$

---

polynomial $u(ch)$ in (6.33) is formally replaced by a new one, say $\bar{u}(ch)$, having the same degree s. The polynomial $w(\tau)$ remains unaffected, provided that $u(h)$ is replaced by $\bar{u}(h)$. Using a Gauss-Legendre formula $(\bar{c}_i, \bar{b}_i)$ based at k abscissae, we arrive at defining the EQUIP(k, s) method, which is described by Algorithm 6.3. The following result may be readily established, on the basis of Theorem 6.3.

Theorem 6.4. *Let y_1 be the approximation to $y(h)$, solution of (6.1), provided by the EQUIP(k, s) method illustrated in Algorithm 6.3. Then, for all $k \geqslant s$,*

$$y_1 - y(h) = O(h^{2s+1}).$$

All quadratic invariants are exactly preserved and, moreover,

$$H(y_1) - H(y_0) = \begin{cases} 0, & \text{if H is a polynomial of degree} \\ & \text{not larger than $2k/s$,} \\ O(h^{2k+1}), & \text{otherwise.} \end{cases}$$

As an example of application, let us consider the Kepler problem described in Section 2.5, with initial conditions (4.43) and eccentricity $e = 0.6$. We recall that the corresponding solution is periodic with period $T = 2\pi$. We compare

TABLE 6.3: Numerical results for the Kepler problem.

			HBVM$(2,2)$				
h	error	rate	$\|H - H_0\|$	rate	$\|M - M_0\|$	$\|A_2 - A_{2,0}\|$	rate
$T/20$	6.3547e-01	–	1.7771e-02	–	4.7518e-14	2.1400e-01	–
$T/40$	1.6956e-01	1.91	4.0167e-04	5.47	4.1078e-14	1.7713e-02	3.59
$T/80$	1.4679e-02	3.53	2.2448e-05	4.16	2.1094e-14	1.1574e-03	3.94
$T/160$	9.7651e-04	3.91	1.5014e-06	3.90	1.9984e-15	7.3515e-05	3.98
$T/320$	6.1901e-05	3.98	9.5283e-08	3.98	3.9968e-15	4.6138e-06	3.99
$T/640$	3.8822e-06	4.00	5.9775e-09	3.99	2.7756e-15	2.8866e-07	4.00
			EQUIP$(12,2)$				
h	error	rate	$\|H - H_0\|$	rate	$\|M - M_0\|$	$\|A_2 - A_{2,0}\|$	rate
$T/20$	3.0426e-01	–	4.4409e-16	–	4.2455e-13	3.1926e-01	–
$T/40$	2.0619e-02	3.88	6.6613e-16	–	5.5511e-15	2.5434e-02	3.65
$T/80$	1.3325e-03	3.95	4.4409e-16	–	9.9920e-16	1.6673e-03	3.93
$T/160$	8.3858e-05	3.99	6.6613e-16	–	2.7756e-15	1.0538e-04	3.98
$T/320$	5.2499e-06	4.00	4.4409e-16	–	2.8866e-15	6.6042e-06	4.00
$T/640$	3.2825e-07	4.00	6.6613e-16	–	3.1086e-15	4.1305e-07	4.00

the HBVM$(2,2)$ (i.e. the 2-stage Gauss) and the EQUIP$(12,2)$ methods with stepsize $h = T/n$, for increasing values of n, to cover 5 periods. [10] In Table 6.3 we list the errors in the solution and in the invariants: as one may see, the latter method is fourth-order, according to Theorem 6.4, whereas the error in the Hamiltonian H, and in the quadratic angular momentum M, is negligible. Conversely, the error for the second component of the Lenz vector, A_2, decreases with the same order as that of the method. As is expected, the conservation of two invariants during the integration procedure results in a more accurate solution than that computed by the 2-stage Gauss method, which preserves only the angular momentum.

We end this section by mentioning that the approach used for conserving only the Hamiltonian can be extended to handle the conservation of multiple invariants. As an example, for two invariants, one could use, assuming that $s \geqslant 3$, a perturbed matrix X_s in the form (compare with (6.29)-(6.30)):

$$X_s(\alpha_1, \alpha_2) := X_s - \alpha_1(e_2 e_1^\top - e_1 e_2^\top) - \alpha_2(e_3 e_2^\top - e_2 e_3^\top),$$

where the two parameters α_1 and α_2 are determined, at each step, in order to enforce the conservation of both invariants. Clearly, this is achieved by using, as before, a suitable line integral.

---

[10]For smaller values of k, say $k = 8$ or $k = 10$, the integrals (6.36) are not adequately approximated, when using the coarsest stepsize $T/20 \approx 0.314$ for the EQUIP$(k,2)$ method, thus resulting in a non-conserved numerical Hamiltonian.

Algorithm 6.4 - HBVM solution of problem (6.42)

$$
\begin{aligned}
&\text{for}\quad i = 1,\ldots,N: \\
&\qquad \text{solve}\quad \hat{\gamma}^i = \mathcal{P}_s^\top \Omega \otimes J\,\nabla H\left(\mathbf{1}\otimes y_{i-1} + h\mathcal{I}_s \otimes I_{2m}\hat{\gamma}^i\right) \\
&\qquad \text{set}\quad y_i := y_{i-1} + h\hat{\gamma}_0^i \\
&\text{such that:}\quad g(y_0, y_N) = 0
\end{aligned}
$$

6.4 Hamiltonian Boundary Value Problems

In this section we shall deal with the use of energy-conserving methods, in the HBVMs class, for solving Hamiltonian problems in the form

$$
\dot{y} = J\nabla H(y), \quad t \in [0,T], \quad g(y(0), y(T)) = 0 \in \mathbb{R}^{2m}, \quad y \in \mathbb{R}^{2m}, \quad (6.42)
$$

where g is a general nonlinear function, satisfying suitable conditions, in order for the problem to be well defined.[11] As an example, if $g(y(0), y(T)) = y(0) - y_0$, one would obtain the *initial value problem* (6.1). In general, g functionally relates the value of $y(0)$ and $y(T)$, so that we speak about a Hamiltonian *two-point boundary value problem*. Moreover, the arguments that we shall present here extend in a quite natural way to the case of *multi-point* boundary value problems for which, given a suitable set of points $0 = t_0 < t_1 < \cdots < t_\nu = T$, the condition is in the form $g(y(t_0), y(t_1), \ldots, y(t_\nu)) = 0 \in \mathbb{R}^{2m}$.

A more "technical" case, which we shall not tackle here, is that of devising periodic solutions, i.e., when the function g in (6.42) is in the form

$$
g(y(0), y(T)) \equiv y(T) - y(0) = 0.
$$

In such a case, a generic (not necessarily Hamiltonian) problem always admits an infinite number of solutions: the phase of the solution is always undetermined, meaning that y_0 could be any point on the given periodic orbit. For Hamiltonian systems, the conservation of energy adds a further source of indeterminacy which reflects the fact that periodic orbits are not isolated. Other sources of indeterminacies may arise in presence of additional first integrals. Consequently, the handling of this case is a much more delicate issue (see the bibliographical notes).

To solve (6.42) numerically, let us consider a suitable *mesh* on the interval

---

[11]I.e., the problem admits a solution, which is also unique in a suitable neighborhood of the solution itself.

$[0, T]$ which, for sake of simplicity, we assume uniform:[12]

$$t_i := ih, \qquad i = 0, \ldots, N, \qquad \text{where} \qquad h = \frac{T}{N}. \qquad (6.43)$$

Denoting, as usual, by y_i the approximation of $y(t_i)$, the application of a HBVM(k, s) method for solving (6.42) is described in Algorithm 6.4, where we have set

$$\hat{\gamma}^i := \begin{pmatrix} \hat{\gamma}_0^i \\ \vdots \\ \hat{\gamma}_{s-1}^i \end{pmatrix} \in \mathbb{R}^{2ms}, \qquad i = 1, \ldots, N,$$

the vector with the coefficients of the piecewise polynomial approximation defined by the method, when restricted to the subinterval $[t_{i-1}, t_i]$:

$$u(t_{i-1} + ch) := y_{i-1} + h \sum_{j=0}^{s-1} \int_0^c P_j(x) \mathrm{d}x \, \hat{\gamma}_j^i, \qquad c \in [0, 1].$$

Consequently, we have to solve the problem *globally*, rather than in a step-by-step fashion. This is better illustrated by casting the discrete problem in vector form. In fact, by defining the vector of the unknowns as

$$\boldsymbol{y} = \left(\, y_0^\top, \ (\hat{\gamma}^1)^\top, \ y_1^\top, \ \ldots, \ (\hat{\gamma}^N)^\top, \ y_N^\top \, \right)^\top \in \mathbb{R}^{2mN(s+1)+2m}, \qquad (6.44)$$

one obtains that the implementation of Algorithm 6.4 is formally equivalent to the solution of the nonlinear system

$$\Phi_N(\boldsymbol{y}) := \begin{pmatrix} g(y_0, y_N) \\ \hat{\gamma}^1 - \mathcal{P}_s^\top \Omega \otimes J \nabla H \left(\mathbf{1} \otimes y_0 + h \mathcal{I}_s \otimes I_{2m} \hat{\gamma}^1 \right) \\ y_1 - y_0 - h \hat{\gamma}_0^1 \\ \vdots \\ y_{N-1} - y_{N-2} - h \hat{\gamma}_0^{N-1} \\ \hat{\gamma}^N - \mathcal{P}_s^\top \Omega \otimes J \nabla H \left(\mathbf{1} \otimes y_{N-1} + h \mathcal{I}_s \otimes I_{2m} \hat{\gamma}^N \right) \\ y_N - y_{N-1} - h \hat{\gamma}_0^N \end{pmatrix} = \mathbf{0}, \qquad (6.45)$$

where $\mathcal{P}_s$, $\mathcal{I}_s$, and Ω are the usual matrices defining a HBVM(k, s) method (see (3.33)–(3.35) on page 95). Provided that problem (6.42) is well defined, the following results may be proved.

Theorem 6.5. *With reference to (6.42)–(6.45), the discrete solution provided by a HBVM(k, s) method satisfies:*

$$\max_{i=1,\ldots,N} \| y_i - y(t_i) \| = O(h^{2s}),$$

---

[12]Clearly, non-uniform meshes could be also considered, to obtain a more efficient computational algorithm.

(that is, the method has order 2s) and, moreover,

$$\max_{i=1,\dots,N} |H(y_i) - H(y_0)| = \begin{cases} 0, & \text{if } H \text{ is a polynomial of degree} \\ & \text{not larger than } 2k/s, \\ O(h^{2k}), & \text{otherwise.} \end{cases}$$

Consequently, even for non-polynomial Hamiltonian functions, by choosing k large enough, we can make the discrete solution to lie on the same energy level set, within round-off errors.

A popular way to solve problem (6.45) is the use of a simplified Newton iteration. In more details, by setting

$$G_0 := \frac{\partial}{\partial y_0} g(y_0, y_N), \qquad G_1 := \frac{\partial}{\partial y_N} g(y_0, y_N), \qquad (6.46)$$

$$\nabla^2 H_i := \nabla^2 H\left(\frac{y_{i-1} + y_i}{2}\right), \qquad i = 1, \dots, N,$$

and considering that

$$\mathcal{P}_s^\top \Omega 1 = e_1 \in \mathbb{R}^s, \qquad \mathcal{P}_s^\top \Omega \mathcal{I}_s = X_s,$$

one obtains the iteration

$$\text{solve:} \quad M_N^\ell \Delta^\ell = -\Phi_N(\boldsymbol{y}^\ell) \qquad (6.47)$$

$$\text{set:} \quad \boldsymbol{y}^{\ell+1} = \boldsymbol{y}^\ell + \Delta^\ell, \qquad \ell = 0, 1, \dots,$$

where matrix M_N^ℓ is defined as

$$M_N^\ell := A_N^\ell - (B_N \otimes J) D^\ell, \qquad (6.48)$$

with

$$A_N^\ell := \qquad (6.49)$$

$$\begin{pmatrix} G_0^\ell & & & & & & & G_1^\ell \\ \boldsymbol{0} & I_s \otimes I_{2m} & \boldsymbol{0} & & & & & \\ -I_{2m} & -he_1^\top \otimes I_{2m} & I_{2m} & & & & & \\ & & \boldsymbol{0} & I_s \otimes I_{2m} & \boldsymbol{0} & & & \\ & & -I_{2m} & -he_1^\top \otimes I_{2m} & I_{2m} & & & \\ & & & & & \ddots & & \\ & & & & & \boldsymbol{0} & I_s \otimes I_{2m} & \boldsymbol{0} \\ & & & & & -I_{2m} & -he_1^\top \otimes I_{2m} & I_{2m} \end{pmatrix},$$

$$
B_N := \begin{pmatrix}
0 & & & \cdots & \cdots & & & & 0 \\
e_1 & hX_s & \mathbf{0} & & & & & & \\
0 & \mathbf{0}^\top & 0 & & & & & & \\
& & e_1 & hX_s & \mathbf{0} & & & & \\
& & 0 & \mathbf{0}^\top & 0 & & & & \\
& & & & & \ddots & & & \\
& & & & & & e_1 & hX_s & \mathbf{0} \\
& & & & & & 0 & \mathbf{0}^\top & 0
\end{pmatrix}, \qquad (6.50)
$$

$$
D^\ell := \begin{pmatrix}
I_{s+1} \otimes \nabla^2 H_1^\ell & & & \\
& \ddots & & \\
& & I_{s+1} \otimes \nabla^2 H_N^\ell & \\
& & & O_{2m}
\end{pmatrix}, \qquad (6.51)
$$

the superscript ℓ meaning that the information at step ℓ is used for evaluating the corresponding functions.

We observe that, from the structure of the matrix B_N, some of its nonzero entries (i.e., the blocks hX_s) are multiplied by the stepsize h while others (i.e., the blocks e_1) are not. The former entries then become negligible, as $h \to 0$, with respect to the latter ones. This fact would in principle allow to further simplify the nonlinear iteration, by retaining, in the coefficient matrix, only the terms which are independent of h. In such a case, matrix M_N^ℓ in (6.47) simplifies to:

$$
M_N^\ell := \qquad\qquad\qquad\qquad\qquad\qquad\qquad\qquad\qquad\qquad (6.52)
$$

$$
\begin{pmatrix}
G_0^\ell & & & & & & G_1^\ell \\
-e_1 \otimes J\nabla^2 H_1^\ell & I_s \otimes I_{2m} & \mathbf{0} & & & & \\
-I_{2m} & -he_1^\top \otimes I_{2m} & I_{2m} & & & & \\
& & & \ddots & & & \\
& & & & -e_1 \otimes J\nabla^2 H_N^\ell & I_s \otimes I_{2m} & \mathbf{0} \\
& & & & -I_{2m} & -he_1^\top \otimes I_{2m} & I_{2m}
\end{pmatrix}.
$$

As an intermediate choice, one may also retain the diagonal term $-\frac{1}{2}h\nabla^2 H_i$ for the update of $\hat{\gamma}_0^i$, which means that we are retaining the symmetric part of matrix hX_s in (6.50), while neglecting its skew-symmetric part (see (3.37) on page 96 for the definition of X_s). This, in turn, is equivalent to perform a Newton-like iteration only on the entries

$$
y_0, \hat{\gamma}_0^1, y_1, \hat{\gamma}_0^2, y_2, \dots, y_{N-1}, \hat{\gamma}_0^N, y_N \qquad (6.53)
$$

TABLE 6.4: Numerical results for the Hill problem, HBVM(k, 1), $N = 50$.

$k \cdot T$	0.1	1.4	2.5	3
	4	5	5	5
1	7	13	15	23
	3.4396e-12	4.3958e-10	3.7935e-10	7.5719e-09
	4	5	5	6
3	7	13	15	24
	2.8096e-16	0	1.4040e-16	1.4040e-16

of the unknown vector (6.44), while performing a fixed-point iteration on the remaining ones.

Remark

It is worth noting that matrix M_N^ℓ defined by (6.48)–(6.51) has a particular sparsity structure, which is usually referred to as *Bordered Almost Block Diagonal (BABD)*. Moreover, in case of *separated boundary conditions*, i.e., of the form

$$g(y_0, y_N) \equiv \left(\begin{array}{c} g_0(y_0) \\ g_1(y_N) \end{array} \right) = 0,$$

with $g_0 \in \mathbb{R}^\mu$ and $g_1 \in \mathbb{R}^{2m-\mu}$, the nonzero rows in G_0 and G_1 are the first μ and the last $m - \mu$, respectively (see (6.46)). In such a case, the BABD structure simplifies, after a suitable permutation of the equations, to an *Almost Block Diagonal (ABD)* one. Even though we shall not study here the solution of ABD and BABD linear systems, it must be stressed that there exist quite efficient algorithms for their solution, which exploit the particular sparsity pattern of the coefficient matrices (see the bibliographical notes).

As an example, let us consider the application of the HBVM(k, 1) method, with $N = 50$, for solving the Hill problem described in Section 2.6.2 (see page 73), which is defined by the Hamiltonian:

$$H(q, p) = p_1 q_2 - p_2 q_1 + \frac{1}{2} p^\top p - \frac{1}{\sqrt{q^\top q}} - q_1^2 + \frac{1}{2} q_2^2. \tag{6.54}$$

In particular, we want to compute the trajectory joining the points [13]

$$q(0) := \left(\begin{array}{cc} 3^{-\frac{1}{3}}, & 0 \end{array} \right)^\top, \qquad q(T) := q(0) + \left(\begin{array}{cc} 0.005, & 0.0044 \end{array} \right)^\top, \tag{6.55}$$

---

[13]We recall that we are using normalized units: one unit of length corresponds to about $\mu^{1/3} R \approx 2.16707 \cdot 10^6$ km (see Section 2.6.2 on page 73).

TABLE 6.5: Numerical results for the optimal control Hill problem, HBVM($k, 1$), $N = 50$.

$k \diagdown T$	0.1	2.1	4.1	6.1
	4	6	5	5
1	8	16	20	27
	3.8528e-09	8.4980e-09	3.2171e-08	6.9807e-08
	4	6	6	6
3	8	15	19	26
	3.9411e-16	6.9226e-17	5.6920e-17	3.8868e-17

with the following final times:

$$T \in \{0.1, \ 1.4, \ 2.5, \ 3\}. \tag{6.56}$$

In Table 6.4, for each combination of (k, T), we list the following three values:

1. the number of iterations (6.47) for solving the discrete problem by using the original matrix M_n^ℓ defined by (6.48)–(6.51);

2. the number of iterations (6.47) for solving the discrete problem by using the simplified matrix M_n^ℓ defined at (6.52);[14]

3. the maximum difference in the numerical Hamiltonian along the computed trajectory.

As one may see, the Hamiltonian error provided by the HBVM($3, 1$) method is more favorable than that provided by the HBVM($1, 1$) method (i.e., the implicit midpoint method). Finally, in Figure 6.2 is the plot of the computed trajectories for the different values of T, as specified in (6.56), when using the HBVM($3, 1$) method.

As a further example, let us consider the application of the HBVM($k, 1$) method, with $N = 50$, for solving the optimal control Hill problem, described in Section 2.6.3 on page 75, which is defined by the Hamiltonian (see (2.48)):

$$H(y, \lambda) = \lambda_1 p_1 + \lambda_2 p_2 + \lambda_3 \left(2p_2 - \frac{q_1}{(q_1^2 + q_2^2)^{\frac{3}{2}}} + 3q_1 - \lambda_3 \right)$$
$$+ \frac{\lambda_3^2 + \lambda_4^2}{2} - \lambda_4 \left(2p_1 + \frac{q_2}{(q_1^2 + q_2^2)^{\frac{3}{2}}} + \lambda_4 \right), \tag{6.57}$$

where $y = \left(\begin{array}{cccc} q_1, & q_2, & p_1, & p_2 \end{array} \right)^\top$ is the vector of the *state variables* and

---

[14]In the case of a HBVM($k, 1$) method (i.e., $s = 1$), the matrix of the intermediate approach, equivalent to performing the Newton iteration only on the entries (6.53), clearly coincides with the original matrix M_n^ℓ defined by (6.48)–(6.51).

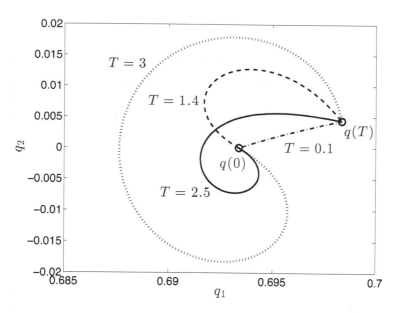

FIGURE 6.2: Numerical solution of problem (6.54)–(6.56) by using the HBVM(3, 1) method with stepsize $h = T/50$.

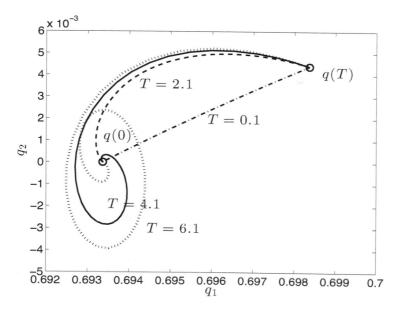

FIGURE 6.3: Numerical solution of problem (6.57)–(6.59) by using the HBVM(3, 1) method with stepsize $h = T/50$.

$\lambda = \begin{pmatrix} \lambda_1, & \lambda_2, & \lambda_3, & \lambda_4 \end{pmatrix}^\top$ is the vector of the *costate variables*. In particular, we compute the trajectories joining the points

$$q(0) \ := \ \begin{pmatrix} 3^{-\frac{1}{3}}, & 0 \end{pmatrix}^\top, \qquad q(T) \ := \ q(0) + \begin{pmatrix} 0.005, & 0.0044 \end{pmatrix}^\top,$$
$$p(0) \ = \ p(T) \ = \ \begin{pmatrix} 0, & 0 \end{pmatrix}^\top, \tag{6.58}$$

when the final times are as follows:

$$T \in \{0.1,\, 2.1,\, 4.1,\, 6.1\}. \tag{6.59}$$

In Table 6.5, for each combination of (k, T), we list the following three values:

1. the number of iterations (6.47) for solving the discrete problem by using the original matrix M_n^ℓ defined by (6.48)–(6.51);

2. the number of iterations (6.47) for solving the discrete problem by using the simplified matrix M_n^ℓ defined at (6.52);

3. the maximum difference in the numerical Hamiltonian along the computed trajectory.

Again, we see that the Hamiltonian error provided by the HBVM$(3,1)$ method is more favorable than that provided by the HBVM$(1,1)$ method. Finally, in Figure 6.3 is the plot of the computed trajectories in the q_1-q_2 plane, for the different values of T, as specified in (6.59), when using the HBVM$(3,1)$ method.

Bibliographical notes

The material of Section 6.1 is adapted from [55, 56]. The methods described in Section 6.2 are adapted from [30, 31]; energy-conserving methods for Poisson systems are described in [25, 75]. The quadratic-invariants and energy-conserving methods described in Section 6.3 are adapted from [47] (see also [32, 42]).

Concerning Hamiltonian boundary value problems, studied in Section 6.4, a classical reference for boundary value problems for ODEs is the book [11]. We refer to [7] (see also [6]), where the delicate case of periodic boundary conditions is discussed in detail and, moreover, selected applications to astrodynamics are described. The optimal control Hill problem has been taken from [99]. The efficient solution of ABD and BABD linear systems is studied in [8, 9].

At last, we mention that multistep extensions of HBVMs have been considered in [46] and low-rank Runge-Kutta methods for stochastic ODEs have been studied in [58].

Appendix A

Auxiliary Material

In this appendix we report a few properties of Legendre polynomials along with a brief description of the Matlab® software implementing the methods. The software is available at the book homepage:
http://web.math.unifi.it/users/brugnano/LIMbook/

A.1   Legendre polynomials

In this section, we collect some properties of shifted Legendre polynomials $\{P_j\}_{j\geqslant 0}$, scaled in order to be orthonormal on the interval $[0,1]$:

$$\deg P_i = i, \qquad \int_0^1 P_i(x)P_j(x)\mathrm{d}x = \delta_{ij}, \qquad \forall i,j = 0,1,\ldots.$$

For their proof see any book on special functions (e.g., [1], see also [82]).

P1. Generalized Rodrigues' formula: for all $n = 0,1,\ldots$, $P_n(x)$ can be defined as

$$P_n(x) = \frac{\sqrt{2n+1}}{n!}\frac{\mathrm{d}^n}{\mathrm{d}x^n}\left[(x^2 - x)^n\right].$$

P2. Gauss-Legendre quadrature: the Gauss-Legendre abscissae

$$0 < c_1 < \cdots < c_s < 1,$$

of the Gauss-Legendre formula of order $2s$, are the zeros of $P_s(x)$. They are symmetrically distributed in the interval $[0,1]$:

$$c_i = 1 - c_{s-i+1}, \qquad i = 1,\ldots,s.$$

The corresponding weights are given by:

$$b_i = \frac{4(2s-1)c_i(1-c_i)}{[sP_{s-1}(c_i)]^2}, \qquad i = 1,\ldots,s,$$

which are positive and, because of the symmetry of the abscissae and **P6** below, satisfy:

$$b_i = b_{s-i+1}, \qquad i = 1, \ldots, s.$$

The formula has order $2s$ and is exact for polynomials of degree $2s - 1$.

P3. Lobatto quadrature: the Lobatto abscissae $\{c_i\}$,

$$0 = c_0 < c_1 < \cdots < c_{s-1} < c_s = 1,$$

of the Lobatto formula of order $2s$, are the zeros of the polynomial

$$(x^2 - x)P_s'(x),$$

where $P_s'(x)$ denotes the derivative of $P_s(x)$. The corresponding weights are given by:

$$b_i = \frac{2s + 1}{s(s + 1)[P_s(c_i)]^2}, \qquad i = 0, 1, \ldots, s,$$

which are, therefore, all positive. Also in this case, one has:

$$c_i = 1 - c_{s-i+1}, \qquad b_i = b_{s-i}, \qquad i = 0, \ldots, s.$$

The formula has order $2s$ and is exact for polynomials of degree $2s - 1$.

P4. Recurrence formula. By setting

$$P_0(x) \equiv 1, \qquad \text{and} \qquad P_1(x) = \sqrt{3}(2x - 1),$$

for $n = 1, 2, \ldots$, one has:

$$P_{n+1}(x) = \frac{2n + 1}{n + 1}\sqrt{\frac{2n + 3}{2n + 1}}(2x - 1)P_n(x) - \frac{n}{n + 1}\sqrt{\frac{2n + 3}{2n - 1}}P_{n-1}(x).$$

P5. Explicit formula:

$$P_n(x) = \sqrt{2n + 1} \sum_{i=0}^{n} \binom{n}{i}\binom{n+i}{i}(-1)^{n+i}x^i, \qquad n = 0, 1, \ldots.$$

P6. Symmetry:

$$P_n(1 - x) = (-1)^n P_n(x), \qquad n = 0, 1, \ldots.$$

P7. Symmetry at the end-points:

$$P_n(0) = (-1)^n \sqrt{2n+1}, \quad P_n(1) = \sqrt{2n+1}, \qquad n = 0, 1, \dots.$$

P8. Derivatives:

$$P_n(x) = \frac{1}{2} \frac{d}{dx} \left[\frac{P_{n+1}(x)}{\sqrt{4(n+1)^2 - 1}} - \frac{P_{n-1}(x)}{\sqrt{4n^2 - 1}} \right], \qquad n = 1, 2, \dots.$$

P9. Integrals:

$$\int_0^x P_0(t)dt = \frac{1}{2} \left[\frac{P_1(x)}{\sqrt{3}} + P_0(x) \right],$$

$$\int_0^x P_n(t)dt = \frac{1}{2} \left[\frac{P_{n+1}(x)}{\sqrt{4(n+1)^2 - 1}} - \frac{P_{n-1}(x)}{\sqrt{4n^2 - 1}} \right], \qquad n = 1, 2, \dots.$$

P10. Shifted Legendre differential equations. The shifted Legendre polynomials satisfy the second order differential equation:

$$\frac{d}{dx} \left[(x^2 - x)P_n'(x) \right] + n(n+1)P_n(x) = 0, \qquad n = 0, 1, \dots.$$

P11. From **P3** and **P10**, it follows that, if $c_0, \dots, c_s$ are the Lobatto abscissae of the formula of order $2s$ (i.e., exact for polynomials of degree $2s - 1$), then

$$\int_0^{c_i} P_s(x)\, dx = 0, \qquad i = 0, 1, \dots, s.$$

P12. A few examples:

$$
\begin{aligned}
P_0(x) &\equiv 1, \\
P_1(x) &= \sqrt{3}\,(2x - 1), \\
P_2(x) &= \sqrt{5}\,(6x^2 - 6x + 1), \\
P_3(x) &= \sqrt{7}\,(20x^3 - 30x^2 + 12x - 1), \\
P_4(x) &= \sqrt{9}\,(70x^4 - 140x^3 + 90x^2 - 20x + 1), \\
P_5(x) &= \sqrt{11}\,(252x^5 - 630x^4 + 560x^3 - 210x^2 + 30x - 1).
\end{aligned}
$$

A.2   Matlab software

In this section we briefly describe a Matlab[1] function, which has been used to perform most of the numerical tests contained in the book. The function, named `hbvm`, in turn relies on a set of Matlab functions, collected as the `HBVMlib1` library, providing the basic building blocks for implementing line integral methods. The function is avaliable at the *book homepage*

http://web.math.unifi.it/users/brugnano/LIMbook/

and has the following format:

[y,t,itot,ierr] = hbvm(sflag,fun,y0,k,s,h,n,nob,stride)

A description of the I/O parameters follows.

Input parameters:

sflag -   flag set to:

- 0, for problems in the form

$$y' = J\nabla H(y), \quad \text{with} \quad y = \begin{pmatrix} q \\ p \end{pmatrix} \text{ and } J = \begin{pmatrix} & I \\ -I & \end{pmatrix}, \quad \text{(A.1)}$$

  and $H(y)$ the Hamiltonian of the problem;

- different from 0, for problems in the form

$$q'' = M\nabla U(q), \quad \text{with} \quad M = M^\top, \quad \text{(A.2)}$$

  and Hamiltonian $H(q,p) = \frac{1}{2}p^\top M p - U(q)$.

fun -   identifier of a function in the form:

- `fun(t,y)` -   evaluates the right-hand side of the differential equation in the form (A.1);
- `fun(y)` -   evaluates the Hamiltonian $H(y)$ in the form (A.1);
- `fun(t,q)` -   evaluates the right-hand side of the differential equation in the form (A.2);
- `fun([q,p])` -   evaluates the Hamiltonian $H(q,p)$ in the form (A.2);
- `fun()` -   evaluates the (symmetric) matrix M in the form (A.2).

---

[1]Matlab® is a popular computing environment produced and distributed by Math-Works.

y0 - initial point of the trajectory, y0 = [q0, p0].

k,s - parameters of the HBVM(k, s) method, $k \geqslant s$. For $k = s$ one has the s-stage Gauss-Legendre method.

h - time step.

n - number of integration steps.

nob - optional flag, assuming the following values (the default value is 0):

- 0 or void ([]), when the blended iteration is used, with numerically evaluated $J\nabla^2 H$ or $\nabla^2 U$;

- function identifier, when the blended iteration is used, with $J\nabla^2 H$ or $\nabla^2 U$ evaluated by the specified function, which has to be in the form G = jaco(t,y) or G = jaco(t,q);

- 1, when a fixed-point iteration is used for solving the generated discrete problems.

stride - on output, the computed solution is returned after each stride steps (the default value is 1).

Output parameters:

y,t - computed solution and mesh (t is optional).

itot - total number of nonlinear iterations (optional).

ierr - error flag (optional) assuming values:

- 0, if the integration procedure ends without problems;

- $\nu > 0$, if at the specified number of steps, convergence problems occurred in the nonlinear iteration;

- -1, if NaNs occurred. In such a case, the integration is stopped at the current step.

Moreover, a warning message is issued in case the specified value of k is not appropriate to provide a practical energy conservation, for the considered stepsize h.

The above function has been used to perform most of the numerical tests contained in the book, the only exception being the use of variable stepsize, the solution of Hamiltonian PDEs, multiple invariants, and Hamiltonian BVPs.

Moreover, in the *Matlab Software* section of the above mentioned webpage one may find the Matlab functions of the problems, as well as a function providing the data to recover the Runge-Kutta form of HBVMs.

Bibliography

[1] M. Abramovitz, I.A. Stegun. *Handbook of Mathematical Functions.* Dover Publications, Inc., New York, 1965.

[2] L. Aceto, D. Trigiante. Symmetric schemes, time reversal symmetry and conservative methods for Hamiltonian systems. *J. Comput. Appl. Math.* **107** (1999) 257–274.

[3] D.S. Alexander, F. Iavernaro, A. Rosa. *Early days in complex dynamics. A history of complex dynamics in one variable during 1906–1942.* History of Mathematics, 38. American Mathematical Society, Providence, RI; London Mathematical Society, London, 2012.

[4] V.I. Arnold, V.V. Kozlov, A.I. Neishtadt. *Mathematical aspects of classical and celestial mechanics.* Dynamical systems III. Third edition. Encyclopaedia of Mathematical Sciences, 3. Springer-Verlag, Berlin, 2006.

[5] P. Amodio, L. Brugnano. A Note on the Efficient Implementation of Implicit Methods for ODEs. *J. Comput. Appl. Math.* **87** (1997) 1–9.

[6] P. Amodio, L. Brugnano, F. Iavernaro. Energy Conservation in the Numerical Solution of Hamiltonian Boundary Value Problems. *AIP Conf. Proc.* **1558** (2013) 35–38.

[7] P. Amodio, L. Brugnano, F. Iavernaro. Energy-conserving methods for Hamiltonian boundary value problems and applications in astrodynamics. *Adv. Comput. Math.* DOI:10.1007/s10444-014-9390-z

[8] P. Amodio, J.R. Cash, G. Fairweather, I. Gladwell, G.L. Kraut, G. Roussos, M. Paprzycki, R.W. Wright. Almost block diagonal linear systems: sequential and parallel solution techniques, and applications. *Numer. Linear Algebra Appl.* **7** (2000) 275–317.

[9] P. Amodio, G. Romanazzi. Algorithm 859: BABDCR – a Fortran 90 package for the solution of Bordered ABD linear systems. *ACM Trans. Math. Softw.* **32** (2006) 597–608.

[10] P. Amodio, I. Sgura. High-order finite difference schemes for the solution of second-order BVPs. *J. Comput. Appl. Math.* **176** (2005) 59–76.

[11] U.M. Ascher, R.M.M. Mattheij, R.D. Russell. *Numerical solution of boundary value problems for ordinary differential equations.* Classics in Applied Mathematics, vol. 13. *Society for Industrial and Applied Mathematics (SIAM)*, Philadelphia, 1995.

[12] J. Barrow-Green. *Poincaré and the three body problem.* History of Mathematics, 11. American Mathematical Society, Providence, RI; London Mathematical Society, London, 1997.

[13] G. Benettin. The elements of Hamiltonian perturbation theory. In: *Hamiltonian systems and Fourier analysis. New prospects for gravitational dynamics,* eds. D. Benest, C. Froeschlé, E. Lega, *Advances in Astronomy and Astrophysics,* Cambridge Scientific Publishers (2005) 1–98.

[14] G. Benettin, A. Giorgilli. On the Hamiltonian interpolation of near to the identity symplectic mappings with application to symplectic integration algorithms. *J. Statist. Phys.* **74** (1994) 1117–1143.

[15] G.P. Berman, F.M. Izrailev. The Fermi-Pasta-Ulam problem: fifty years of progress. *Chaos* **15**, 1 (2005) 015104, 18 pp.

[16] P. Betsch, P. Steinmann. Inherently Energy Conserving Time Finite Elements for Classical Mechanics. *J. Comp. Phys.* **160** (2000) 88–116.

[17] P. Betsch, P. Steinmann. Conservation properties of a time FE method. I. Time-stepping schemes for N-body problems. *Internat. J. Numer. Methods Engrg.* **49**, 5 (2000) 599–638.

[18] C.L. Bottasso. A new look at finite elements in time: a variational interpretation of Runge-Kutta methods. *Appl. Numer. Math.* **25** (1997) 355–368.

[19] J.P. Boyd. *Chebyshev and Fourier spectral methods. Second edition.* Dover Publications, Inc., Mineola, NY, 2001.

[20] T.J. Bridges. Multisymplectic structures and wave propagation. *Math. Proc. Cambridge Philos. Soc.* **121** (1997) 147–190.

[21] T.J. Bridges, S. Reich. Multi-symplectic integrators: numerical schemes for Hamiltonian PDEs that conserve symplecticity. *Physics Letters A* **284** (2001) 184–193.

[22] T.J. Bridges, S. Reich. Multi-symplectic spectral discretizations for the Zakharov-Kuznetsov and shallow water equations. *Physica D* **152** (2001) 491–504.

[23] T.J. Bridges, S. Reich. Numerical methods for Hamiltonian PDEs. *J. Phys. A: Math. Gen.* **39** (2006) 5287–5320.

[24] L. Brugnano. Blended Block BVMs (B3VMs): A Family of Economical Implicit Methods for ODEs. *J. Comput. Appl. Math.* **116** (2000) 41–62.

[25] L. Brugnano, M. Calvo, J.I. Montijano, L. Rández. Energy preserving methods for Poisson systems. *J. Comput. Appl. Math.* **236** (2012) 3890–3904.

[26] L. Brugnano, G. Frasca Caccia, F. Iavernaro. Efficient implementation of Gauss collocation and Hamiltonian Boundary Value Methods. *Numer. Algor.* **65** (2014) 633–650.

[27] L. Brugnano, G. Frasca Caccia, F. Iavernaro. Efficient implementation of geometric integrators for separable Hamiltonian problems. *AIP Conference Proceedings* **1558** (2013) 734–737.

[28] L. Brugnano, G. Frasca Caccia, F. Iavernaro. Hamiltonian Boundary Value Methods (HBVMs) and their efficient implementation. *Mathematics in Engineering, Science and Aerospace* **5**, 4 (2014) 343–411.

[29] L. Brugnano, G. Frasca Caccia, F. Iavernaro. Energy conservation issues in the numerical solution of the semilinear wave equation. *Preprint*, 2014 `arXiv:1410.7009[math.NA]`.

[30] L. Brugnano, F. Iavernaro. Line Integral Methods which preserve all invariants of conservative problems. *J. Comput. Appl. Math.* **236** (2012) 3905–3919.

[31] L. Brugnano, F. Iavernaro. Recent Advances in the Numerical Solution of Conservative Problems. *AIP Conference Proc.* **1493** (2012) 175–182.

[32] L. Brugnano, F. Iavernaro. Geometric Integration by Playing with Matrices. *AIP Conference Proceedings* **1479** (2012) 16–19.

[33] L. Brugnano, F. Iavernaro, C. Magherini. Efficient implementation of Radau collocation methods. *Appl. Numer. Math.* **87** (2015) 100–113.

[34] L. Brugnano, F. Iavernaro. *Line integral methods and their application to the numerical solution of conservative problems*, Lecture Notes, 2013. `arXiv:1301.2367[math.NA]`

[35] L. Brugnano, F. Iavernaro, T. Susca. Hamiltonian BVMs (HBVMs): implementation details and applications. *AIP Conf. Proc.* **1168** (2009) 723–726.

[36] L. Brugnano, F. Iavernaro, T. Susca. Numerical comparisons between Gauss-Legendre methods and Hamiltonian BVMs defined over Gauss points. *Monografias de la Real Academia de Ciencias de Zaragoza* **33** (2010) 95–112.

[37] L. Brugnano, F. Iavernaro, D. Trigiante. Analysis of Hamiltonian Boundary Value Methods (HBVMs) for the numerical solution of polynomial Hamiltonian dynamical systems. (2009) `arXiv:0909.5659v1[math.NA]` (published as reference [48]).

[38] L. Brugnano, F. Iavernaro, D. Trigiante. Hamiltonian BVMs (HBVMs): a family of "drift-free" methods for integrating polynomial Hamiltonian systems. *AIP Conf. Proc.* **1168** (2009) 715–718.

[39] L. Brugnano, F. Iavernaro, D. Trigiante. *The Hamiltonian BVMs (HBVMs) Homepage*, 2010. `arXiv:1002.2757`

[40] L. Brugnano, F. Iavernaro, D. Trigiante. Hamiltonian Boundary Value Methods (Energy Preserving Discrete Line Methods). *JNAIAM J. Numer. Anal. Ind. Appl. Math.* **5**, 1-2 (2010) 17–37.

[41] L. Brugnano, F. Iavernaro, D. Trigiante. Numerical Solution of ODEs and the Columbus' Egg: Three Simple Ideas for Three Difficult Prob-

lems. *Mathematics in Engineering, Science and Aerospace* **1**, 4 (2010) 407–426.

[42] L. Brugnano, F. Iavernaro, D. Trigiante. Energy and quadratic invariants preserving integrators of Gaussian type. *AIP Conf. Proc.* **1281** (2010) 227–230.

[43] L. Brugnano, F. Iavernaro, D. Trigiante. A note on the efficient implementation of Hamiltonian BVMs. *J. Comput. Appl. Math.* **236** (2011) 375–383.

[44] L. Brugnano, F. Iavernaro, D. Trigiante. The Lack of Continuity and the Role of Infinite and Infinitesimal in Numerical Methods for ODEs: the Case of Symplecticity. *Appl. Math. Comput.* **218** (2012) 8053–8063.

[45] L. Brugnano, F. Iavernaro, D. Trigiante. A simple framework for the derivation and analysis of effective one-step methods for ODEs. *Appl. Math. Comput.* **218** (2012) 8475–8485.

[46] L. Brugnano, F. Iavernaro, D. Trigiante. A two-step, fourth-order method with energy preserving properties. *Comp. Phys. Comm.* **183** (2012) 1860–1868.

[47] L. Brugnano, F. Iavernaro, D. Trigiante. Energy and QUadratic Invariants Preserving integrators based upon Gauss collocation formulae. *SIAM J. Numer. Anal.* **50**, 6 (2012) 2897–2916.

[48] L. Brugnano, F. Iavernaro, D. Trigiante. Analisys of Hamiltonian Boundary Value Methods (HBVMs): a class of energy-preserving Runge-Kutta methods for the numerical solution of polynomial Hamiltonian systems. *Communications in Nonlinear Science and Numerical Simulation* **20** (2015) 650–667.

[49] L. Brugnano, C. Magherini. Blended Implementation of Block Implicit Methods for ODEs. *Appl. Numer. Math.* **42** (2002) 29–45.

[50] L. Brugnano, C. Magherini. The BiM Code for the Numerical Solution of ODEs. *J. Comput. Appl. Math.* **164-165** (2004) 145–158.

[51] L. Brugnano, C. Magherini, F. Mugnai. Blended Implicit Methods for the Numerical Solution of DAE Problems. *J. Comput. Appl. Math.* **189** (2006) 34–50.

[52] L. Brugnano, C. Magherini. Some Linear Algebra Issues Concerning the Implementation of Blended Implicit Methods. *Numer. Lin. Alg. Appl.* **12** (2005) 305–314.

[53] L. Brugnano, C. Magherini. Recent Advances in Linear Analysis of Convergence for Splittings for Solving ODE problems. *Appl. Numer. Math.* **59** (2009) 542–557.

[54] L. Brugnano, C. Magherini, F. Mugnai. Blended Implicit Methods for the Numerical Solution of DAE Problems. *J. Comput. Appl. Math.* **189** (2006) 34–50.

[55] L. Brugnano, Y. Sun. Enhanced HBVMs for the numerical solution of Hamiltonian problems with multiple invariants. *AIP Conference Proc.* **1588** (2013) 754–757.

[56] L. Brugnano, Y. Sun. Multiple invariants conserving Runge-Kutta type methods for Hamiltonian problems. *Numer. Algor.* **65** (2014) 611–632.

[57] L. Brugnano, D. Trigiante. *Solving ODEs by Linear Multistep Initial and Boundary Value Methods*, Gordon and Breach, Amsterdam, 1998.

[58] K. Burrage, P.M. Burrage. Low rank Runge-Kutta methods, symplecticity and stochastic Hamiltonian problems with additive noise. *J. Comput. Appl. Math.* **236** (2012) 3920–3930.

[59] K. Burrage, J.C. Butcher. Stability criteria for implicit Runge-Kutta methods. *SIAM J. Numer. Anal.* **16** (1979) 46–57.

[60] J.C. Butcher. Implicit Runge-Kutta processes. *Math. Comput.* **18** (1964) 50–64.

[61] J.C. Butcher. On the implementation of implicit Runge-Kutta methods. *BIT* **16** (1976) 237–240.

[62] J.C. Butcher. Numerical methods for ordinary differential equations in the 20th century. *J. Comput. Appl. Math.* **125** (2000) 1–29.

[63] M. Calvo, M.P. Laburta, J.I. Montijano, L. Rández. Error growth in the numerical integration of periodic orbits. *Math. Comput. Simulation* **81** (2011) 2646–2661.

[64] M.P. Calvo, E. Hairer. Accurate long-term integration of dynamical systems. *Appl. Numer. Math.* **18** (1995) 95-105.

[65] B. Cano. Conserved quantities of some Hamiltonian wave equations after full discretization. *Numer. Math.* **103** (2006) 197–223.

[66] C. Canuto, M.Y. Hussaini, A. Quarteroni, T.A. Zang. *Spectral Methods in Fluid Dynamics*. Springer-Verlag, New York, 1988.

[67] E. Celledoni, V. Grimm, R.I. McLachlan, D.I. McLaren, D. O'Neale, B. Owren, G.R.W. Quispel. Preserving energy resp. dissipation in numerical PDEs using the "average vector field" method. *J. Comput. Phys.* **231**, 20 (2012) 6770–6789.

[68] E. Celledoni, R.I. McLachlan, D.I. McLaren, B. Owren, G.R.W. Quispel, W.M. Wright. Energy-preserving Runge-Kutta methods. *M2AN Math. Model. Numer. Anal.* **43**, 4 (2009) 645–649.

[69] E. Celledoni, R.I. McLachlan, B. Owren, G.R.W. Quispel. Energy-Preserving Integrators and the Structure of B-series. *Found. Comput. Math.* **10** (2010) 673–693.

[70] A. Celletti. *Perturbation Theory in Celestial Mechanics*. Encyclopedia of Complexity and System Science, R.A. Meyer ed., Springer, 2009.

[71] F. Ceschino, J. Kuntzmann. *Problémes Différentiels de Conditions Initiales*. Dunod, Paris, 1963.

[72] P. Channell. Symplectic integration algorithms. *Tech. Report Report AT-6ATN 83-9, Los Alamos National Laboratory*, 1983.

[73] P. Chartier, E. Faou, A. Murua. An algebraic approach to invariant preserving integrators: the case of quadratic and Hamiltonian invariants. *Numer. Math.* **103**, 4 (2006) 575–590.

[74] J.B. Chen, M.Z. Qin. Multi-symplectic Fourier pseudospectral method for the nonlinear Schrödinger equation. *Electron. Trans. Numer. Anal.* **12** (2001) 193–204.

[75] D. Cohen, E. Hairer. Linear energy-preserving integrators for Poisson systems. *BIT* **51**, 1 (2011) 91–101.

[76] D. Cohen, E. Hairer, C. Lubich. Conservation of energy, momentum and actions in numerical discretizations of non-linear wave equations. *Numer. Math.* **110** (2008) 113–143.

[77] P. Console, E. Hairer. Reducing round-off errors in symmetric multistep methods. *Jour. Comput. Appl. Math.* **262** (2014) 217–222.

[78] G. Contopoulos. On the existence of a third integral of motion. *Astron. J.* **68** (1963) 1–14.

[79] G. Contopoulos. A classification of the integrals of motion. *Astrophys. J.* **138** (1963) 1297–1305.

[80] M. Crouzeix. Sur la B-stabilité des méthodes de Runge-Kutta. *Numer. Math.* **32** (1979) 75–82.

[81] G. Dahlquist, Å. Bijörk. *Numerical Methods*, Prentice-Hall, Englewood Cliffs, N.J., 1974.

[82] P.J. Davis, P. Rabinowitz. *Methods of Numerical Integration, second edition*. Academic Press, Inc., Orlando, FL, 1984.

[83] F. Diacu, P. Holmes. *Celestial encounters. The origins of chaos and stability.* Princeton University Press, Princeton, NJ, 1996.

[84] E. Faou. *Geometric numerical integration and Schrödinger equations.* Zurich Lectures in Advanced Mathematics. European Mathematical Society (EMS), Zürich, 2012.

[85] A. Farrés, J. Laskar, S. Blanes, F. Casas, J. Makazaga, A. Murua. High precision symplectic integrators for the Solar System. *Celest. Mech. Dyn. Astr.* **116** (2013) 141-174.

[86] K. Feng. On Difference Schemes and Symplectic Geometry. In *Proceedings of the 1984 Beijing symposium on differential geometry and differential equations.* Science Press, Beijing, 1985, pp. 42–58.

[87] K. Feng, M. Quin. *Symplectic Geometric Algorithms for Hamiltonian Systems.* Springer, Zhejiang Publishing United Group Zhejiang Science and Technology Publishing House, 2010.

[88] E. Fermi, J.R. Pasta, S. Ulam. *Studies of nonlinear problems, I.* Los Alamos report LA-1940. Los Alamos Scientific Laboratory, 1955.

[89] T. Flå. A numerical energy conserving method for the DNLS equation. *J. Comput. Phys.* **101** (1992) 71–79.

[90] B. Forneberg, G.B. Whitham. A Numerical and Theoretical Study of Certain Nonlinear Wave Phenomena. *Proc. R. Soc. Lond. A* **289** (1978) 373–403.

[91] J. Frank. Conservation of wave action under multisymplectic discretizations. *J. Phys. A: Math. Gen.* **39** (2006) 5479–5493.

[92] J. Frank, B.E. Moore, S. Reich. Linear PDEs and Numerical Methods that Preserve a Multisymplectic Conservation Law. *SIAM J. Sci. Comput.* **28** (2006) 260–277.

[93] G. Frasca Caccia. *A new efficient implementation for HBVMs and their application to the semilinear wave equation.* PhD Thesis, Dipartimento di Matematica e Informatica "U. Dini", University of Firenze, Italy, 2015. (http://web.math.unifi.it/users/brugnano/LIMbook/data/PhDthesisGFC.pdf)

[94] J. de Frutos, T. Ortega, J.M. Sanz-Serna. A Hamiltonian, explicit algorithm with spectral accuracy for the "good" Boussinesq system. *Comput. Methods Appl. Mech. Engrg.* **80** (1990) 417–423.

[95] D. Furihata. Finite-difference schemes for nonlinear wave equation that inherit energy conservation property. *J. Comput. Appl. Math.* **134**, 1-2 (2001) 37–57.

[96] D. Furihata, T. Matsuo. *Discrete variational derivative method. A structure-preserving numerical method for partial differential equations.* CRC Press, Boca Raton, FL, 2011.

[97] L. Galgani, A. Giorgilli, A. Martinoli, S. Vanzini. On the problem of energy equipartition for large systems of the FermiPastaUlam type: analytical and numerical estimates. *Physica D* **59** (1992) 334–348.

[98] Z. Ge, J.E. Marsden. Lie-Poisson Hamilton-Jacobi theory and Lie-Poisson integrators. *Phys. Lett. A* **133** (1988) 134–139.

[99] V.M. Guibout, D.J. Scheeres. Solving two-point boundary value problems using the Hamilton-Jacobi theory. *Proceedings of the 2nd WSEAS Int. Conference on Applied and Theoretical Mechanics*, Venice, Italy, November 20–22, 2006, pp. 174–182.

[100] H. Goldstein, C.P. Poole, J.L. Safko. *Classical Mechanics.* Addison Wesley, 2001.

[101] O. Gonzales. Time integration and discrete Hamiltonian systems. *J. Nonlinear Sci.* **6** (1996) 449–467.

[102] W. Gröbner. *Gruppi, Anelli e Algebre di Lie.* Collana di Informazione Scientifica "Poliedro", Edizioni Cremonese, Rome, 1975.

[103] E. Hairer. Energy preserving variant of collocation methods. *JNAIAM J. Numer. Anal. Ind. Appl. Math.* **5**, 1-2 (2010) 73–84.

[104] E. Hairer, C. Lubich. Spectral semi-discretisations of weakly nonlinear wave equations over long times. *Found. Comput. Math.* **8** (2008) 319–334.

[105] E. Hairer, C. Lubich, G. Wanner. *Geometric Numerical Integration. Structure-Preserving Algorithms for Ordinary Differential Equations*, Second ed., Springer, Berlin, 2006.

[106] E. Hairer, G. Wanner. Algebraically stable and implementable Runge-Kutta methods of high order. *SIAM J. Numer. Anal.* **18** (1981) 1098–1108.

[107] E. Hairer, G. Wanner. *Solving Ordinary Differential Equations II. Stiff and Differential-Algebraic Problems, 2nd edition.* Springer-Verlag, Berlin, 1996.

[108] E. Hairer, C.J. Zbinden. On conjugate symplecticity of B-series integrators. *IMA J. Numer. Anal.* (2012) 1–23.

[109] B. Hasselblatt, A. Katok. The development of dynamics in the 20th century and the contribution of Jürgen Moser. *Ergodic Theory Dynam. Systems* **22**, 5 (2002) 1343–1364.

[110] M. Hénon, C. Heiles. The Applicability of the Third Integral Of Motion: Some Numerical Experiments. *The Astrophysical Journal* **69** (1964) 73–79.

[111] B.M. Herbst, M.J. Ablowitz. Numerical chaos, symplectic integrators, and exponentially small splitting distances. *J. Comput. Phys.* **105**, 1 (1993) 122–132.

[112] P.J. van der Houwen, J.J.B. de Swart. Triangularly implicit iteration methods for ODE-IVP solvers. *SIAM J. Sci. Comput.* **18** (1997) 41–55.

[113] P.J. van der Houwen, J.J.B. de Swart. Parallel linear system solvers for Runge-Kutta methods. *Adv. Comput. Math.* **7**, 1-2 (1997) 157–181.

[114] W. Hu, Z. Deng, S. Han, W. Zhang. Generalized multi-symplectic integrators for a class of Hamiltonian nonlinear wave PDEs. *J. Comput. Phys.* **235** (2013) 394–406.

[115] M. Huang. A Hamiltonian approximation to simulate solitary waves of the Kortweg-de Vries equation. *Math. Comp.* **56**, 194 (1991) 607–620.

[116] B.L. Hulme. One-Step Piecewise Polynomial Galerkin Methods for Initial Value Problems. *Math. Comp.*, **26**, 118 (1972) 415–426.

[117] B.L. Hulme. Discrete Galerkin and related one-step methods for ordinary differential equations. *Math. Comp.* **26** (1972) 881–891.

[118] F. Iavernaro, B. Pace. s-Stage Trapezoidal Methods for the Conservation of Hamiltonian Functions of Polynomial Type. *AIP Conf. Proc.* **936** (2007) 603–606.

[119] F. Iavernaro, B. Pace. Conservative Block-Boundary Value Methods for the Solution of Polynomial Hamiltonian Systems. *AIP Conf. Proc.* **1048** (2008) 888–891.

[120] F. Iavernaro, D. Trigiante. High-order symmetric schemes for the energy conservation of polynomial Hamiltonian problems. *JNAIAM J. Numer. Anal. Ind. Appl. Math.* **4**,1-2 (2009) 87–101.

[121] A.L. Islas, C.M. Schober. On the preservation of phase space structure under multisymplectic discretization. *J. Comput. Phys.* **197**,2 (2004) 585–609.

[122] A.L. Islas, C.M. Schober. Backward error analysis for multisymplectic discretizations of Hamiltonian PDEs. *Math. Comput. Simulation* **69** (2005) 290–303.

[123] A.L. Islas, C.M. Schober. Conservation properties of multisymplectic integrators. *Future Generation Computer Systems* **22** (2006) 412–422.

[124] C.G.J. Jacobi. Sur le movement d'un point et sur un cas particulier du problème des trois corps. *Comptes Rendus de l'Académie des Sciences de Paris* **3** (1836) 59–61.

[125] S. Jiménez. Derivation of the discrete conservation laws for a family of finite difference schemes. *Appl. Math. Comput.* **64** (1994) 13–45.

[126] C. Kane, J.E. Marsden, M. Ortiz. Symplectic-energy-momentum preserving variational integrators. *J. Math. Phys.* **40**,7 (1999) 3353–3371.

[127] S. Koide, D. Furihata. Nonlinear and linear conservative finite difference schemes for regularized long wave equation. *Japan J. Indust. Appl. Math.* **26**,1 (2009) 15–40.

[128] A. Kurganov, J. Rauch. The Order of Accuracy of Quadrature Formulae for Periodic Functions. *Advances in Phase Space Analysis of Partial Differential Equations*, A. Bove et al. (eds.), Birkhäuser, Boston, 2009.

[129] M.P. Laburta, J.I. Montijano, L. Rández, M. Calvo. Numerical methods for non conservative perturbations of conservative problems. *Comput. Phys. Commun.* **187** (2015) 72–82.

[130] V. Lakshmikantham, D. Trigiante. *Theory of Difference Equations. Numerical Methods and Applications*. Academic Press, 1988.

[131] F.M. Lasagni. Canonical Runge-Kutta methods. *ZAMP* **39** (1988) 952-953.

[132] R.I. Mc Lachlan, G.R.W. Quispel, N. Robidoux. Geometric integration using discrete gradient. *Phil. Trans. R. Soc. Lond. A* **357** (1999) 1021–1045.

[133] B. Leimkulher, S. Reich. *Simulating Hamiltonian Dynamics*. Cambridge University Press, Cambridge, 2004.

[134] C.W. Li, M.Z. Qin. A symplectic difference scheme for the infinite-dimensional Hamilton system. *J. Comput. Math.* **6** (1988) 164–174.

[135] S. Li, L. Vu-Quoc. Finite difference calculus invariant structure of a class of algorithms for the nonlinear Klein-Gordon equation. *SIAM J. Numer. Anal.* **32** (1995) 1839–1875.

[136] X. Lu, R. Schmid. A symplectic algorithm for wave equations. *Math. and Comput. in Simulat.* **43** (1997) 29–38.

[137] C. Lubich. *From Quantum to Classical Molecular Dynamics: Reduced Models and Numerical Analysis.* European Mathematical Solciety, 2008.

[138] J.E. Marsden, G.P. Patrick, S. Shkoller. Multi-symplectic geometry, variational integrators, and nonlinear PDEs. *Comm. Math. Phys.* **199** (1999) 351–395.

[139] J.E. Marsden, J.M. Wendlandt. Mechanical Systems with Symmetry, Variational Principles, and Integration Algorithms. in *"Current and Future Directions in Applied Mathematics"* M. Alber, B. Hu, and J. Rosenthal, Eds., Birkhäuser, 1997, pp. 219–261.

[140] T. Matsuo. New conservative schemes with discrete variational derivatives for nonlinear wave equations. *J. Comput. Appl. Math.* **203** (2007) 32–56.

[141] T. Matsuo, M. Sugihara, D. Furihata, M. Mori. Spatially accurate dissipative or conservative finite difference schemes derived by the discrete variational method. *Japan J. Indust. Appl. Math.* **19**, 3 (2002) 311–330.

[142] C.R. Menyuk. Some properties of the discrete Hamiltonian method. *Phys. D* **11**, 1-2 (1984) 109–129.

[143] K.R. Meyer, G.R. Hall, D. Offin. *Introduction to Hamiltonian dynamical systems and the N-body problem.* Applied Mathematical Sciences, vol. 90, 2nd edn. Springer, New York (2009).

[144] B. Moore, S. Reich. Backward error analysis for multi-symplectic integration methods. *Numer. Math.* **95** (2003) 625–652.

[145] A.C. Newell. *Solitons in mathematics and physics.* CBMS-NSF Regional Conference Series in Applied Mathematics **48**, SIAM, Philadelphia, PA, 1985.

[146] M. Oliver, M. West C. Wulff. Approximate momentum conservation for spatial semidiscretization of semilinear wave equations. *Numer. Math.* **97** (2004) 493–535.

[147] M.A. Porter, N.J. Zabusky, B. Hu, D.K. Campbell. Fermi, Pasta, Ulam and the birth of experimental mathematics. *American Scientist* **97**, 3 (2009) 214–221.

[148] M.Z. Qin, M.Q. Zhang. Multi-stage symplectic schemes of two kinds of Hamiltonian systems for wave equations. *Computer Math. Appl.* **19**, 10 (1990) 51–62.

[149] G.R.W. Quispel, D.I. McLaren. A new class of energy-preserving numerical integration methods. *J. Phys. A: Math. Theor.* **41** (2008) 045206 (7pp).

[150] D.C. Rapaport. *The Art of Molecular Dynamics Simulation, 2nd edition.* Cambridge University Press, Cambridge, 2004.

[151] R.D. Ruth. A canonical integration technique. *IEEE Trans. Nucl. Sci.* **30** (1983) 2669–2671.

[152] J.M. Sanz-Serna. Runge-Kutta schemes for Hamiltonian systems. *BIT* **28** (1988) 877–883.

[153] J.M. Sanz-Serna, M.P. Calvo. *Numerical Hamiltonian Problems.* Chapman & Hall, London, 1994.

[154] T. Schlick. *Molecular Modeling and Simulation: An Interdisciplinary Guide, 2nd edition,* Springer, New York, 2010.

[155] Z.J. Shang. KAM theorem of symplectic algorithms for Hamiltonian systems. *Numer. Math.* **83** (1999) 477–496.

[156] J.C. Simo, N. Tarnow. A new energy and momentum conserving algorithm for the non-linear dynamics of shells. *Internat. J. Numer. Methods Engrg.* **37** (1994) 2527–2549.

[157] J.C. Simo, N. Tarnow and K.K. Wong. Exact energy-momentum conserving algorithms and symplectic schemes for nonlinear dynamics. *Comput. Methods Appl. Mech. Engrg.* **100** (1992) 63–116.

[158] V. Simoncini. *Computational methods for linear matrix equations.* 2014. (http://www.dm.unibo.it/~simoncin/public_matrixeq_rev.pdf)

[159] W. Strauss, L. Vázquez. Numerical solution of a nonlinear Klein-Gordon equation. *J. Comput. Phys.* **28** (1978) 271–278.

[160] R.D. Skeel. Symplectic integration with floating-point arithmetic and other approximations. *Appl. Numer. Math.* **29** (1999) 3–18.

[161] A.M. Stuart, A.R. Humphries. *Dynamical systems and numerical analysis.* Cambridge Monographs on Applied and Computational Mathematics, 2. Cambridge University Press, Cambridge, 1996.

[162] Y.B. Suris. On the canonicity of mappings that can be generated by methods of Runge-Kutta type for integrating systems $x'' = \partial U / \partial x$. *U.S.S.R. Comput. Math. and Math. Phys.* **29**, 1 (1989) 138–144.

[163] Y.B. Suris. *The problem of integrable discretization: Hamiltonian approach.* Progress in Mathematics, 219. Birkhäuser Verlag, Basel, 2003.

[164] M. Suzuki. Fractal decomposition of exponential operators with applications to many-body theories and Monte Carlo simulations. *Phys. Lett. A* **146** (1990) 319–323.

[165] Q. Tang, C.M. Chen. Continuous finite element methods for Hamiltonian systems. *Appl. Math. Mech.* **28**, 8 (2007) 1071–1080.

[166] W. Tang, Y. Sun. Time finite element methods: a unified framework for numerical discretizations of ODEs. *Appl. Math. Comput.* **219**, 4 (2012) 2158–2179.

[167] R. de Vogelaere. Methods of integration which preserve the contact transformation property of Hamiltonian equations. *Tech. Report No 4, Dept. Mathem., Univ. of Notre Dame, Notre Dame, Ind.,* 1956.

[168] M. Valtonen, H. Karttunen. *The three-body problem.* Cambridge University Press, Cambridge, 2006.

[169] J. Wang. A note on multisymplectic Fourier pseudospectral discretization for the nonlinear Schrödinger equation. *Appl. Math. Comput.* **191** (2007) 31–41.

[170] D. Wang, A. Xiao, X. Li. Parametric symplectic partitioned Runge-Kutta methods with energy-preserving properties for Hamiltonian systems. *Comp. Phys. Comm.* **184** (2013) 303–310.

[171] J.A.C. Weideman. Numerical Integration of Periodic Functions: A Few Examples. *Amer. Math. Monthly* **109**, 1 (2002) 21–36.

[172] S.B. Wineberg, J.F. Mc Grath, E.F. Gabl, L.R. Scott, C.E. Southwell. Implicit spectral methods for wave propagation problems. *J. Comp. Physics* **97** (1991) 311–336.

[173] T.H. Wlodarczyk. *Stability and preservation properties of multisymplectic integrators.* PhD thesis, Department of Mathematics in the College of Sciences at the University of Central Florida, Orlando, Florida, 2007. (`http://etd.fcla.edu/CF/CFE0001817/Wlodarczyk_Tomasz_H_200708_PhD.pdf`)

Index

Printed and bound by CPI Group (UK) Ltd, Croydon, CR0 4YY

23/10/2024

01778242-0009